Statistical Mechanics for Chemical Thermodynamics and Kinetics

Andrea Amadei · Massimiliano Aschi

Statistical Mechanics for Chemical Thermodynamics and Kinetics

Springer

Andrea Amadei
Department of Chemical Sciences
and Technology
University of Roma Tor Vergata
Roma, Italy

Massimiliano Aschi
Department of Physical and Chemical
Sciences
University of L'Aquila
Aquila, Italy

ISBN 978-3-031-77931-2 ISBN 978-3-031-77929-9 (eBook)
https://doi.org/10.1007/978-3-031-77929-9

This Springer imprint is published by the registered company Springer Nature Switzerland AG
The registered company address is: Gewerbestrasse 11, 6330 Cham, Switzerland

Preface

Over the last three decades we have witnessed an enormous development in theoretical chemistry. Nowadays it is rather rare to find basic experimental research works in which theoretical–computational data are not used in support of (or for prediction of) experimental data of any nature. Applied research itself, especially in the bio-pharmaceutical and materials development fields, is gaining enormous benefits from the use of methodologies based on models and/or simulations at both the electronic and atomistic levels. The main reason for this clear increase of interest for theoretical models in chemistry is originally linked to the concomitant development of computational chemistry, an ancillary discipline of theoretical chemistry, aided by the enormous technological and algorithmic progress. In fact, chemistry is a complex science in which fully analytical models, even when mathematically elegant, are typically idealized and thus lack a realistic description of the phenomena studied and are unable to produce quantitative results of real interest for interpreting and/or predicting experimental data. Differently, a theoretical–computational model is expected to quantitatively reproduce rather complex observables in the thermodynamic, kinetic and spectroscopic fields, possibly providing an exceptional tool for interpreting and/or predicting experimental data. The present strong interplay between the theoretical developments and computational efficiency opens the way to models specifically designed for the study of complex chemical processes in realistic conditions (reactions in solution, spectroscopic signal of solutes, conformational transitions in macromolecules, etc.) practically inaccessible to theoretical–computational chemistry only a few decades ago. In this context it is obvious that modern theoretical chemistry should necessarily focus on modeling complex molecular systems (i.e., involving a huge number of interacting particles and characterized by the coexistence of quantum and classical-like degrees of freedom) representing the challenge of the forefront research in chemistry. This book, conceived according to such a purpose, is specifically aimed at providing an advanced textbook including most of the fundamental concepts and derivations to be used for the statistical mechanical modeling of the typical complex molecular systems of chemical interest. Starting from the basic principles of quantum mechanics and specifically introducing the concepts and derivations for treating quantum-classical systems (Chap. 1 and the

Appendix), we describe in detail the fundamental principles of statistical mechanics (Chap. 2) and their advanced applications to chemical thermodynamics and kinetics (Chaps. 3 and 4). Some practical examples of the use of the theoretical derivations described combined with computational methods based on molecular dynamics simulations are also shown illustrating the efficiency and accuracy of actual theoretical–computational models for complex molecular systems. Our intention is to report the procedure used, highlighting the intimate link between the theoretical principles and their application with the results obtained without, however, lingering on the technical aspects (the reader is referred to the proper literature for more details on the theoretical–computational methods employed).

The book, essentially reporting the topics of the lessons held by one of the authors (AA) for the Master of Chemistry course at the University of Rome "Tor Vergata", provides a rigorous description of the theoretical principles and derivations addressing the advanced physical–mathematical aspects of the statistical mechanics in chemistry, thus bridging the gap between basic textbooks and more specialized publications. Therefore, it is mainly intended for the Master of Chemistry students and the PhD students in Physical and Theoretical Chemistry, although we believe that also post-doctoral researchers in molecular sciences can benefit from it.

Roma, Italy Andrea Amadei
Aquila, Italy Massimiliano Aschi

Contents

Chapter 1
Mechanics and Dynamics of Quantum Systems

Abstract In this introductory chapter we describe in detail the quantum mechanical concepts and relations to be used for deriving the basic principles of statistical mechanics as well as its advanced applications to realistic molecular systems. After introducing the time-dependent Schroedinger equation, its general analytical solution for stationary Hamiltonians and the Born–Oppenheimer approximation, we specifically address the treatment of molecular systems characterized by the coexistence of quantum and classical-like degrees of freedom (mixed quantum-classical systems).

1.1 General Considerations and the Schroedinger Equation

The mechanics of quantum systems can be achieved once a suitable expression for each relevant observable is available and in quantum mechanics this is accomplished by assuming that all the definitions and relations present for classical observables be the same for the corresponding quantum observables (classical analogy). Only the spin-related observables, without any classical analog, need then to be expressed by ex novo derivations. The dynamics of quantum systems can be addressed on the basis of the equivalence between the system dynamical state and a proper time-dependent quantum state. In particular, any dynamical state characterized by some observables being invariant, i.e., their measured values are well defined and time-independent, can be considered as corresponding at any time to a given eigenstate of such observables, which hence must be commuting observables (see the Appendix). In order to address the time dependence of the quantum state of a system we need that a completely general equation of *motion* for quantum states, playing the same basic role of Newton's equations in (non-relativistic) classical mechanics, be available. Given its general nature we should be able to derive such kind of equations even by considering the simplest quantum mechanical system conceivable: a free spinless particle of mass m whose quantum state $|\Psi(t)\rangle$ should be given by a time-dependent quantum state always corresponding to an eigenstate of its linear momentum (i.e., eigenstate of the momentum components), as it follows from the fact that its dynamical state corresponds in classical terms to a uniform motion with hence a well

A. Amadei and M. Aschi, *Statistical Mechanics for Chemical Thermodynamics and Kinetics*, https://doi.org/10.1007/978-3-031-77929-9_1

defined constant linear momentum vector (this also implies that the three components of the linear momentum must be commuting observables). For the simple system we are considering, we can write the Hamiltonian operator

$$\widehat{H} = mc^2\sqrt{1 + \frac{\widehat{\boldsymbol{p}}^2}{m^2c^2}} \tag{1.1}$$

where $\widehat{\boldsymbol{p}}$ is the (Cartesian) linear momentum (vector) operator and $\widehat{H}$ is the Hamiltonian operator we defined by classical analogy from the relativistic Hamiltonian of a classical free particle, corresponding only to the particle kinetic energy, given by $H = mc^2\sqrt{1 + \frac{\boldsymbol{p}^2}{m^2c^2}}$. In case the momentum eigenstate, as usual, corresponds to a low particle velocity compared to light speed, we can use the simplified Hamiltonian $H \cong mc^2 + \frac{\boldsymbol{p}^2}{2m}$ provided by $\frac{\boldsymbol{p}^2}{m^2c^2} \cong 0$ to obtain

$$\widehat{H} = mc^2\sqrt{1 + \frac{\widehat{\boldsymbol{p}}^2}{m^2c^2}} \cong mc^2 + \frac{\widehat{\boldsymbol{p}}^2}{2m} \tag{1.2}$$

Note that the relativistic Hamiltonian in the $\frac{\boldsymbol{p}^2}{m^2c^2} \to 0$ limit (non-relativistic limit) coincides with the non-relativistic Hamiltonian except for the constant mc^2 which is irrelevant in Newton's equations. Therefore, such a constant can be taken as a reference energy value which can be subtracted from the relativistic Hamiltonian in order to recover exactly the non-relativistic expression. It is also worth to note that for a free particle $\widehat{H}$ can be expressed as a series of powers of the observable $\widehat{\boldsymbol{p}}$ (for the exact relativistic expression this is accomplished by Taylor expansion) thus providing $\widehat{\boldsymbol{p}}\widehat{H} - \widehat{H}\widehat{\boldsymbol{p}} = 0$ and hence any $\widehat{\boldsymbol{p}}$ eigenstate must also be a Hamiltonian eigenstate, that is

$$\widehat{\boldsymbol{p}}\,|\Psi(t)\rangle = \boldsymbol{p}\,|\Psi(t)\rangle \tag{1.3}$$

$$\widehat{H}\,|\Psi(t)\rangle = \mathcal{U}\,|\Psi(t)\rangle \tag{1.4}$$

with $\boldsymbol{p}$ and $\mathcal{U}$ the momentum and Hamiltonian eigenvalue, respectively.

Empirical evidence, discovered at the beginning of the twentieth century, showed that any free particle can be also considered as a propagating wave with frequency ν and wavelength λ related to the particle energy and linear momentum (i.e., the particle Hamiltonian and momentum eigenvalues) via the Planck and De Broglie equations

$$h\nu = \mathcal{U} \tag{1.5}$$

$$|\boldsymbol{p}| = \frac{h}{\lambda} \tag{1.6}$$

with h a universal constant introduced by Planck and then named Planck's constant. Therefore, expressing the spinless particle quantum state within the Cartesian coordinates $\boldsymbol{r}$ eigenstate basis set, i.e., $|\Psi(t)\rangle = \Psi(\boldsymbol{r}, t)$, and on the basis of the empirical evidence assuming a simple plane wave propagating in the direction of the invariant linear momentum corresponding to the particle dynamical state, we obtain

$$\Psi(\boldsymbol{r}, t) = Ae^{i(\boldsymbol{k}\cdot\boldsymbol{r}-\omega t)} \tag{1.7}$$

where A is a constant, $\omega = 2\pi\nu$ and $\boldsymbol{k}$ is the wave propagation vector with $|\boldsymbol{k}| = 2\pi/\lambda$ and $\boldsymbol{k}/|\boldsymbol{k}| = \boldsymbol{p}/|\boldsymbol{p}|$. Note that Eq. 1.7 is also the historical reason for naming wavefunctions the quantum states expressed in the coordinate eigenstate basis set, even when not resembling any propagating wave. From Planck and De Broglie equations we can write $\omega = \mathcal{U}/\hbar$, $\hbar\boldsymbol{k} = \boldsymbol{p}$ $(2\pi\hbar = h)$ and hence Eq. 1.7 becomes

$$\Psi(\boldsymbol{r}, t) = Ae^{\frac{i}{\hbar}(\boldsymbol{p}\cdot\boldsymbol{r}-\mathcal{U}t)} \tag{1.8}$$

readily providing

$$-i\hbar\nabla_r\Psi = \boldsymbol{p}\Psi \tag{1.9}$$

$$i\hbar\frac{\partial}{\partial t}\Psi = \mathcal{U}\Psi \tag{1.10}$$

where

$$\nabla_r = \begin{bmatrix} \partial/\partial r_1 \\ \partial/\partial r_2 \\ \partial/\partial r_3 \end{bmatrix}$$

is the Cartesian gradient operator. Finally, by using Eqs. 1.3 and 1.4 into Eqs. 1.9 and 1.10 we obtain

$$\widehat{\boldsymbol{p}} = -i\hbar\nabla_r \tag{1.11}$$

$$\widehat{H} = i\hbar\frac{\partial}{\partial t} \tag{1.12}$$

Note that while Eq. 1.11 expresses the momentum operator in the Cartesian coordinate eigenstate basis set, Eq. 1.12 relating the Hamiltonian operator with the time derivative which is clearly independent of the choice of the basis set used must be considered as the completely general form of the Hamiltonian operator when acting on the system time-dependent quantum state (i.e., the system dynamical state). It is worth to remark here that time is not a mechanical property and hence does not correspond to an observable within the meaning used in this context. Therefore, time must be considered as a physical parameter, mathematically corresponding to a scalar, possibly involved in the definition of quantum states and observables.

It is a *postulate* of quantum mechanics that Eqs. 1.11 and 1.12, although obtained by using a simple spinless free particle system, are correct for whatever mechanical system considered (spin particles, interacting particles, particles subjected to an external field, etc.). We can therefore write the completely general Schroedinger equation

$$i\hbar\frac{\partial}{\partial t}|\Psi(t)\rangle = \widehat{H}\ |\Psi(t)\rangle \tag{1.13}$$

and its bra form

$$-\,i\hbar\frac{\partial}{\partial t}\langle\Psi(t)| = \langle\Psi(t)|\ \widehat{H} \tag{1.14}$$

where now $|\Psi(t)\rangle$ and $\widehat{H}$ represent the time-dependent quantum state and the Hamiltonian operator of any possible mechanical system. Interestingly, from Eqs. 1.13 and 1.14 we have

$$i\hbar\frac{\partial}{\partial t}\Big(\langle\Psi(t)|\Psi(t)\rangle\Big) = -\langle\Psi(t)|\widehat{H}|\Psi(t)\rangle + \langle\Psi(t)|\widehat{H}|\Psi(t)\rangle = 0 \tag{1.15}$$

$$\begin{aligned} i\hbar\frac{\partial}{\partial t}\Big(\langle\Psi(t)|\widehat{H}|\Psi(t)\rangle\Big) &= -\langle\Psi(t)|\widehat{H}^2|\Psi(t)\rangle + i\hbar\langle\Psi(t)|\frac{\partial\widehat{H}}{\partial t}|\Psi(t)\rangle + \langle\Psi(t)|\widehat{H}^2|\Psi(t)\rangle \\ &= i\hbar\langle\Psi(t)|\frac{\partial\widehat{H}}{\partial t}|\Psi(t)\rangle \end{aligned} \tag{1.16}$$

clearly showing that the length of the dynamical state is an invariant regardless of the system and conditions considered and for stationary Hamiltonian systems (i.e., $\partial\widehat{H}/\partial t = 0$) the expectation energy is, as expected, constant in time (i.e., it is also an invariant).

Moreover, from the definition of the momentum operator within the Cartesian coordinate eigenstate basis set we also readily obtain the completely general commutation relation between coordinates and momenta

$$\widehat{p_l}\widehat{r_{l'}} - \widehat{r_{l'}}\widehat{p_l} = -i\hbar\delta_{l,l'} \tag{1.17}$$

where $\widehat{p_l}, \widehat{r_{l'}}$ are the Cartesian components of the particles momentum and position operators.

It is worth to note that for any kind of mechanical system composed by slow particles compared to light, once using for each particle the approximation due to $\frac{\widehat{p}^2}{m^2c^2} \cong 0$ to express its kinetic energy operator, we must have $\widehat{H} \cong Mc^2 + \widehat{H}'$ where $M = \sum_l m_l$ is the total mass of the system given by the summation of the particles masses and $\widehat{H}'$ contains all the terms of the Hamiltonian operator corresponding to the classical non-relativistic kinetic energy and potential energy as well as to the possible spin interaction terms (for spinless particles or spin particles with negligible spin interactions $\widehat{H}'$ thus coincides with the Hamiltonian operator as obtained by classical analogy from the non-relativistic classical Hamiltonian). Therefore, for any system in the non-relativistic limit with Hamiltonian eigenstates $|\Phi\rangle$ and time-dependent quantum state $|\Psi(t)\rangle$ we can write (by using the Hamiltonian eigenvalue equation and the Schroedinger equation with $\widehat{H} \cong Mc^2 + \widehat{H}'$)

$$\widehat{H}'|\Phi\rangle \cong \mathcal{U}'|\Phi\rangle \tag{1.18}$$

$$i\hbar\frac{\partial}{\partial t}|\Psi'(t)\rangle \cong \widehat{H}'\,|\Psi'(t)\rangle \tag{1.19}$$

where $\mathcal{U}' = \mathcal{U} - Mc^2$, $|\Psi'(t)\rangle = e^{\frac{i}{\hbar}Mc^2 t}|\Psi(t)\rangle$ and we used the fact that $e^{\frac{i}{\hbar}Mc^2 t}$ being purely a function of time is a scalar thus commuting with any observable and hence with $\widehat{H}$ and $\widehat{H}'$. These last equations plainly show that for most of the quantum mechanical systems of interest in chemistry where all the relevant particles can be considered slow with respect to light, we can use $\widehat{H}'$ as the proper Hamiltonian operator thus fully disregarding, as in classical mechanics, the constant Mc^2. Finally, it must be noted that $|\Psi'(t)\rangle$ differs from the, in principle, exact time-dependent Schroedinger equation solution $|\Psi(t)\rangle$ by simply a phase factor, thus implying that for any observable of interest $\widehat{A}$ we always have $\langle\Psi'(t)|\widehat{A}|\Psi'(t)\rangle = \langle\Psi(t)|\widehat{A}|\Psi(t)\rangle$ and hence $|\Psi'(t)\rangle$ can be considered as the proper system time-dependent quantum state except for an irrelevant phase factor (note that for any couple of dynamical states of the system $|\Psi_1(t)\rangle$, $|\Psi_2(t)\rangle$ we have $\langle\Psi_1'(t)|\widehat{A}|\Psi_2'(t)\rangle = \langle\Psi_1(t)|\widehat{A}|\Psi_2(t)\rangle$). In the following, dealing with mechanical systems in the non-relativistic limit, we will always consider Eqs. 1.18 and 1.19 as exact, neglecting both the constant Mc^2 and the phase factor $e^{\frac{i}{\hbar}Mc^2 t}$ (i.e., for such systems we can use $\widehat{H} = \widehat{H}'$ and so $\mathcal{U} = \mathcal{U}'$ and $|\Psi(t)\rangle = |\Psi'(t)\rangle$).

1.2 The Solution of the Schroedinger Equation for a Stationary Hamiltonian Operator

From the previous section we can now express the time evolution of a dynamical quantum state of typical physical–chemical systems, by means of the Schroedinger equation (Eq. 1.13) with the Hamiltonian operator and its eigenvalues provided by the classical analogy. By using as discrete orthonormal basis set the Hamiltonian eigenstates to express the (normalized) dynamical quantum state and considering a stationary Hamiltonian with hence stationary eigenstates $|\eta_l\rangle$ and eigenvalues $\mathcal{U}_l$, we can readily solve the Schroedinger equation. In fact, in such a condition using $|\Psi(t)\rangle = \sum_l c_l(t)|\eta_l\rangle$, we obtain from Eq. 1.13

$$i\hbar\sum_l \dot{c}_l|\eta_l\rangle = \sum_l c_l(t)\widehat{H}|\eta_l\rangle = \sum_l \mathcal{U}_l c_l(t)|\eta_l\rangle \tag{1.20}$$

which necessarily implies that each $c_l(t)$ component fulfills the equation

$$i\hbar\dot{c}_l = \mathcal{U}_l c_l(t) \tag{1.21}$$

providing $c_l(t) = c_l(0)e^{-\frac{i}{\hbar}\mathcal{U}_l t}$. Therefore, the general solution of the Schroedinger equation when considering a stationary Hamiltonian operator can be readily obtained via

$$|\Psi(t)\rangle = \sum_l c_l(0)e^{-\frac{i}{\hbar}\mathcal{U}_l t}|\eta_l\rangle \tag{1.22}$$

clearly showing a time-independent expectation energy (in agreement with the general result given by Eq. 1.16)

$$\langle\Psi(t)|\widehat{H}|\Psi(t)\rangle = \sum_l |c_l(0)|^2\mathcal{U}_l \tag{1.23}$$

with

$$\langle\Psi(t)|\Psi(t)\rangle = \sum_l |c_l(0)|^2 = 1 \tag{1.24}$$

Finally, the difference of two solutions

$$|\Delta\Psi(t)\rangle = \sum_l \Delta c_l(0)e^{-\frac{i}{\hbar}\mathcal{U}_l t}|\eta_l\rangle \tag{1.25}$$

obviously determined only by the initial state difference given by all the $\Delta c_l(0)$, provides

$$\langle\Delta\Psi(t)|\Delta\Psi(t)\rangle = \sum_l |\Delta c_l(0)|^2 \tag{1.26}$$

indicating that no attractors can be present in the state space for quantum trajectories of systems with time-independent Hamiltonians.

1.3 The Coordinates-Spin Representation

In chemistry the atomic–molecular level is the proper description of any mechanical system, thus requiring that electrons and nuclei be the relevant particles to be treated as point masses and charges. Such approximation which is excellent for electrons is actually inaccurate when modeling in detail the nuclear magnetic behavior which requires higher order approximations than the point charge one; however, for most of the issues relevant to chemistry, not involving significant effects of nuclei–electrons interactions at very short distances, we can safely consider each nucleus as a point charge neglecting such higher order effects. Therefore, by using the electrons and nuclei Cartesian positions $\hat{\boldsymbol{r}}_e, \hat{\boldsymbol{r}}_n$ and momenta $\hat{\boldsymbol{p}}_e, \hat{\boldsymbol{p}}_n$ to express the Hamiltonian operator we obtain

$$\hat{H} = \hat{K}_n(\hat{\boldsymbol{p}}_n) + \hat{K}_e(\hat{\boldsymbol{p}}_e) + \hat{E}_P(\hat{\boldsymbol{r}}_n, \hat{\boldsymbol{r}}_e) + \hat{V}_{m,s}(\hat{\boldsymbol{s}}_n, \hat{\boldsymbol{s}}_e, \hat{\boldsymbol{p}}_e, \hat{\boldsymbol{r}}_n, \hat{\boldsymbol{r}}_e) \tag{1.27}$$

where $\hat{K}_n$, $\hat{K}_e$ are the kinetic energy operators for nuclei and electrons, respectively, $\hat{E}_P$ is the potential energy operator, $\hat{V}_{m,s}$ provides the magnetic-spin interactions operator and $\hat{\boldsymbol{s}}_n, \hat{\boldsymbol{s}}_e$ are the multidimensional vectors defined by the 3-D spin vector operators $\hat{\boldsymbol{s}}_{n,k}, \hat{\boldsymbol{s}}_{e,k}$ of the nuclei and electrons, respectively

$$\hat{\boldsymbol{s}}_n = \begin{bmatrix} \hat{\boldsymbol{s}}_{n,1} \\ \hat{\boldsymbol{s}}_{n,2} \\ \cdot \\ \hat{\boldsymbol{s}}_{n,k} \\ \cdot \end{bmatrix} \quad \hat{\boldsymbol{s}}_e = \begin{bmatrix} \hat{\boldsymbol{s}}_{e,1} \\ \hat{\boldsymbol{s}}_{e,2} \\ \cdot \\ \hat{\boldsymbol{s}}_{e,k} \\ \cdot \end{bmatrix}$$

with k the nuclei or electrons particle index. Note that $\hat{E}_P$, just like the kinetic energy operators, is fully defined by the classical analogy and hence for point charge particles it is provided simply by the Coulomb electrostatic interactions. Differently, $\hat{V}_{m,s}$ involving the particles magnetic interactions including the spin interactions does not correspond in general to a classical mechanical observable and hence its expression in terms of the spin observables needs to be derived on quantum mechanical ground [1]. However, due to the empirical evidence that each particle spin generates a magnetic dipole we can consider $\hat{V}_{m,s}$ as corresponding to a purely magnetic energy term and hence such an operator can be considered as virtually independent of the nuclei momenta $\hat{\boldsymbol{p}}_n$ as the nuclear velocities are too low to provide a significant current density for magnetic interactions. When using the Cartesian coordinates and the particle spin commuting observables to define, by means of their eigenstates, the basis set (mixed basis set) we can write

$$\hat{H} = \hat{K}_n(\hat{\boldsymbol{p}}_n) + \hat{K}_e(\hat{\boldsymbol{p}}_e) + \hat{E}_P(\boldsymbol{r}_n, \boldsymbol{r}_e) + \hat{V}_{m,s}(\tilde{\boldsymbol{s}}_n, \tilde{\boldsymbol{s}}_e, \hat{\boldsymbol{p}}_e, \boldsymbol{r}_n, \boldsymbol{r}_e) \tag{1.28}$$

$$\hat{\boldsymbol{p}}_n = \begin{bmatrix} \hat{\boldsymbol{p}}_{n,1} \\ \hat{\boldsymbol{p}}_{n,2} \\ \cdot \\ \hat{\boldsymbol{p}}_{n,k} \\ \cdot \\ \hat{\boldsymbol{p}}_{n,N_n} \end{bmatrix} = -i\hbar \begin{bmatrix} \partial/\partial r_{n,1} \\ \partial/\partial r_{n,2} \\ \cdot \\ \cdot \\ \cdot \\ \partial/\partial r_{n,3N_n} \end{bmatrix} \tag{1.29}$$

$$\hat{\boldsymbol{p}}_e = \begin{bmatrix} \hat{\boldsymbol{p}}_{e,1} \\ \hat{\boldsymbol{p}}_{e,2} \\ \cdot \\ \hat{\boldsymbol{p}}_{e,k} \\ \cdot \\ \hat{\boldsymbol{p}}_{e,N_e} \end{bmatrix} = -i\hbar \begin{bmatrix} \partial/\partial r_{e,1} \\ \partial/\partial r_{e,2} \\ \cdot \\ \cdot \\ \cdot \\ \partial/\partial r_{e,3N_e} \end{bmatrix} \tag{1.30}$$

with $\widetilde{s}_n$ and $\widetilde{s}_e$ the multidimensional vectors of the matrices representing the nuclei and electrons spin operators within the spin eigenstates basis set, i.e., for each particle the matrix expressing the z-component of the spin operator is diagonal (see the Appendix), and N_n, N_e the nuclear and electronic total number of particles. Note that each term of the Hamiltonian operator involving the coordinates even if not necessarily expressed as a combination of powers and/or exponentials of the coordinate operators when using the coordinate eigenstates basis set must be given by the same expression substituting the coordinate operators with the corresponding eigenvalues as each Hamiltonian term is an observable itself (see the Appendix). By grouping all the terms involving the electronic coordinates and the spin operators together into the electron-spin Hamiltonian operator $\widehat{H}_{es}$ we obtain

$$\widehat{H} = \widehat{H}_{es} + \widehat{K}_n \tag{1.31}$$

$$\widehat{H}_{es} = \widehat{K}_e + \widehat{E}_P + \widehat{V}_{m,s} \tag{1.32}$$

and hence the Hamiltonian eigenvalue equation can be written as

$$\left(\widehat{H}_{es} + \widehat{K}_n\right) \Phi = \mathcal{U} \Phi \tag{1.33}$$

where $\Phi(\boldsymbol{r}_n, \boldsymbol{r}_e, \boldsymbol{\gamma}_n, \boldsymbol{\gamma}_e)$ is the Hamiltonian eigenstate with eigenvalue $\mathcal{U}$ as expressed in the coordinates-spin eigenstates basis set according to the usual formalism as illustrated in Sect. 5.4 of the Appendix and $\boldsymbol{\gamma}_n, \boldsymbol{\gamma}_e$ are the nuclear and electronic multidimensional vectors given by the single particle spin eigenstate unit vectors.

1.4 The Born–Oppenheimer Approximation

We can proceed further realizing that $\widehat{H}_{es}$, which does not involve $\widehat{\boldsymbol{p}}_n$, commutes with each nuclear coordinate and thus its eigenstates (necessarily antisymmetric quantum states for electrons permutations [1]) must be eigenstates also for the nuclear coordinates. Therefore, we can express the $\widehat{H}_{es}$ eigenstates as $\Phi_{es}(\boldsymbol{r}'_n, \boldsymbol{r}_e, \boldsymbol{\gamma}_n, \boldsymbol{\gamma}_e)\, \delta(\boldsymbol{r}_n - \boldsymbol{r}'_n)$ providing

$$\begin{aligned} &\widehat{H}_{es}(\boldsymbol{r}_n, \boldsymbol{r}_e, \widehat{\boldsymbol{p}}_e, \widetilde{s}_n, \widetilde{s}_e)\, \Phi_{es}(\boldsymbol{r}'_n, \boldsymbol{r}_e, \boldsymbol{\gamma}_n, \boldsymbol{\gamma}_e)\delta(\boldsymbol{r}_n - \boldsymbol{r}'_n) \\ &= \mathcal{U}_{es}(\boldsymbol{r}'_n)\, \Phi_{es}(\boldsymbol{r}'_n, \boldsymbol{r}_e, \boldsymbol{\gamma}_n, \boldsymbol{\gamma}_e)\delta(\boldsymbol{r}_n - \boldsymbol{r}'_n) \end{aligned} \tag{1.34}$$

once Φ_{es} is defined by

$$\widehat{H}_{es}(\boldsymbol{r}'_n, \boldsymbol{r}_e, \widehat{\boldsymbol{p}}_e, \widetilde{s}_n, \widetilde{s}_e)\, \Phi_{es}(\boldsymbol{r}'_n, \boldsymbol{r}_e, \boldsymbol{\gamma}_n, \boldsymbol{\gamma}_e) = \mathcal{U}_{es}(\boldsymbol{r}'_n)\, \Phi_{es}(\boldsymbol{r}'_n, \boldsymbol{r}_e, \boldsymbol{\gamma}_n, \boldsymbol{\gamma}_e) \tag{1.35}$$

and hence it corresponds to the electron-spin Hamiltonian eigenstate with eigenvalue $\mathcal{U}_{es}$, as obtained considering the electron-spin quantum degrees of freedom as embedded into the nuclear electrostatic field provided by the fixed nuclei with positions $\boldsymbol{r}'_n$ (i.e., the nuclear positions are regarded as physical parameters determining the field). Each jth Hamiltonian eigenstate Φ_j (necessarily a finite length quantum state) can then be defined as a linear combination of the $\widehat{H}_{es}$ eigenstates (i.e., using the mixed basis set $\Phi_{es}(\boldsymbol{r}'_n, \boldsymbol{r}_e, \boldsymbol{\gamma}_n, \boldsymbol{\gamma}_e)\, \delta(\boldsymbol{r}_n - \boldsymbol{r}'_n)$) furnishing

$$\begin{aligned}\Phi_j(\boldsymbol{r}_n, \boldsymbol{r}_e, \boldsymbol{\gamma}_n, \boldsymbol{\gamma}_e) &= \sum_l \int w_{j,l}(\boldsymbol{r}'_n)\, \Phi_{es,l}(\boldsymbol{r}'_n, \boldsymbol{r}_e, \boldsymbol{\gamma}_n, \boldsymbol{\gamma}_e)\, \delta(\boldsymbol{r}_n - \boldsymbol{r}'_n)\, d\boldsymbol{r}'_n \\ &= \sum_l w_{j,l}(\boldsymbol{r}_n)\, \Phi_{es,l}(\boldsymbol{r}_n, \boldsymbol{r}_e, \boldsymbol{\gamma}_n, \boldsymbol{\gamma}_e) \end{aligned} \tag{1.36}$$

By inserting Eq. 1.36 into Eq. 1.33 we obtain for the jth Hamiltonian eigenstate

$$\sum_l \mathcal{U}_{es,l} \Phi_{es,l} w_{j,l} + \sum_l \widehat{K}_n \Phi_{es,l} w_{j,l} = \mathcal{U}_j \sum_l \Phi_{es,l} w_{j,l} \tag{1.37}$$

where $\widehat{K}_n$ is given by the non-relativistic kinetic energy expression, providing

$$\widehat{K}_n = -\hbar^2 \sum_{k=1}^{N_n} \frac{\nabla^2_{r_{n,k}}}{2m_{n,k}} \tag{1.38}$$

with $-\hbar^2 \nabla^2_{r_{n,k}} = (-i\hbar\nabla_{r_{n,k}}) \cdot (-i\hbar\nabla_{r_{n,k}})$, $-i\hbar\nabla_{r_{n,k}}$ the 3-D gradient vector operator representing the kth nucleus momenta and $m_{n,k}$ the corresponding nuclear mass. From the definition of the nuclear kinetic energy, we can then write

$$\begin{aligned}\widehat{K}_n w_{j,l} \Phi_{es,l} &= -\sum_{i=1}^{3N_n} \frac{\hbar^2}{2m_{n,i}} \frac{\partial}{\partial r_{n,i}} \left(\frac{\partial w_{j,l}}{\partial r_{n,i}} \Phi_{es,l} + \frac{\partial \Phi_{es,l}}{\partial r_{n,i}} w_{j,l} \right) \\ = &-\sum_{i=1}^{3N_n} \frac{\hbar^2}{2m_{n,i}} \left(\frac{\partial^2 w_{j,l}}{\partial r^2_{n,i}} \Phi_{es,l} + 2\frac{\partial \Phi_{es,l}}{\partial r_{n,i}} \frac{\partial w_{j,l}}{\partial r_{n,i}} + \frac{\partial^2 \Phi_{es,l}}{\partial r^2_{n,i}} w_{j,l} \right) \end{aligned} \tag{1.39}$$

where the index i runs over all the nuclear Cartesian components of the particles momenta. Equation 1.39 can be significantly simplified once considering that only tiny variations of the electronic distribution $|\Phi_{es}|^2$ can be induced by a single nuclear coordinate change and that the nuclear masses are much larger than the electron one. Therefore, from

$$\frac{\hbar^2}{m_{n,i}} \frac{\partial \Phi_{es,l}}{\partial r_{n,i}} \cong 0 \tag{1.40}$$

$$\frac{\hbar^2}{2m_{n,i}} \frac{\partial^2 \Phi_{es,l}}{\partial r^2_{n,i}} \cong 0 \tag{1.41}$$

we obtain the Born–Oppenheimer approximation

$$\widehat{K}_n w_{j,l} \Phi_{es,l} \cong \Phi_{es,l} \widehat{K}_n w_{j,l} \tag{1.42}$$

which, when inserted in Eq. 1.37, provides

$$\sum_l \Phi_{es,l} \left(\widehat{K}_n + \mathcal{U}_{es,l}\right) w_{j,l} \cong \mathcal{U}_j \sum_l \Phi_{es,l} w_{j,l} \tag{1.43}$$

The last equation, which must be fulfilled for any possible electron–nuclear position, necessarily implies for any $w_{j,l}(\boldsymbol{r}_n)$

$$\left(\widehat{K}_n + \mathcal{U}_{es,l}\right) w_{j,l} \cong \mathcal{U}_j w_{j,l} \tag{1.44}$$

and therefore

$$\begin{aligned} &\widehat{H}(\boldsymbol{r}_n, \boldsymbol{r}_e, \widehat{\boldsymbol{p}}_n, \widehat{\boldsymbol{p}}_e, \widetilde{\boldsymbol{s}}_n, \widetilde{\boldsymbol{s}}_e)\ \Phi_{es,l}(\boldsymbol{r}_n, \boldsymbol{r}_e, \boldsymbol{\gamma}_n, \boldsymbol{\gamma}_e)\ w_{j,l}(\boldsymbol{r}_n) \\ \cong\ &\mathcal{U}_j\ \Phi_{es,l}(\boldsymbol{r}_n, \boldsymbol{r}_e, \boldsymbol{\gamma}_n, \boldsymbol{\gamma}_e)\ w_{j,l}(\boldsymbol{r}_n) \end{aligned} \tag{1.45}$$

Equations 1.44 and 1.45 clearly indicate that within the Born–Oppenheimer approximation we can express the Hamiltonian eigenstates Φ_j, with eigenvalues $\mathcal{U}_j$, by combining each lth electron-spin eigenstate $\Phi_{es,l}$ with any of the corresponding nuclear eigenstates $w_{l,k}$ providing

$$\Phi_j \cong \Phi_{l,k}(\boldsymbol{r}_n, \boldsymbol{r}_e, \boldsymbol{\gamma}_n, \boldsymbol{\gamma}_e) = \Phi_{es,l}(\boldsymbol{r}_n, \boldsymbol{r}_e, \boldsymbol{\gamma}_n, \boldsymbol{\gamma}_e)\ w_{l,k}(\boldsymbol{r}_n) \tag{1.46}$$

where $\Phi_{es,l}$ and $w_{l,k}$ are defined by

$$\begin{aligned} &\widehat{H}_{es}(\boldsymbol{r}_n, \boldsymbol{r}_e, \widehat{\boldsymbol{p}}_e, \widetilde{\boldsymbol{s}}_n, \widetilde{\boldsymbol{s}}_e)\ \Phi_{es,l}(\boldsymbol{r}_n, \boldsymbol{r}_e, \boldsymbol{\gamma}_n, \boldsymbol{\gamma}_e) \\ =\ &\mathcal{U}_{es,l}(\boldsymbol{r}_n)\ \Phi_{es,l}(\boldsymbol{r}_n, \boldsymbol{r}_e, \boldsymbol{\gamma}_n, \boldsymbol{\gamma}_e) \end{aligned} \tag{1.47}$$

$$\left[\widehat{K}_n(\widehat{\boldsymbol{p}}_n) + \mathcal{U}_{es,l}(\boldsymbol{r}_n)\right] w_{l,k}(\boldsymbol{r}_n) = \mathcal{U}_{l,k} w_{l,k}(\boldsymbol{r}_n) \tag{1.48}$$

and, clearly, $\Phi_{l,k} = \Phi_{es,l} w_{l,k}$ is the Born-Oppenheimer Hamiltonian eigenstate with eigenvalue $\mathcal{U}_{l,k} \cong \mathcal{U}_j$ fully identified by the l and k indices (i.e., the l, k couple substitutes the previous j index). Note that the Born–Oppenheimer approximation can be inaccurate for degenerate or quasi-degenerate eigenstates, with hence energy differences similar or even smaller than the neglected terms. However, the detailed treatment of such peculiar cases is beyond the scope of the present book and therefore it will not be addressed.

1.5 The Treatment of Mixed Quantum–Classical Systems

We can proceed further by considering from Eq. 1.38

$$\widehat{K}_n = \widehat{\boldsymbol{p}}_n^T \frac{\widetilde{M}_n^{-1}}{2} \widehat{\boldsymbol{p}}_n \tag{1.49}$$

with $\widetilde{M}_n$ the Cartesian (diagonal) mass tensor of the nuclei and clearly

$$\left[\widehat{\boldsymbol{p}}_n^T\right]_k = \left[\widehat{\boldsymbol{p}}_n\right]_k = -i\hbar \frac{\partial}{\partial[\boldsymbol{r}_n]_k} \tag{1.50}$$

Given the analogy of Eq. 1.49 with the kinetic energy classical expression we can consider any arbitrary set of nuclear generalized coordinates $\boldsymbol{q}(\boldsymbol{r}_n)$ as new observables providing

$$\widehat{K}_n = \widehat{\boldsymbol{p}}_n^T \frac{\widetilde{M}_n^{-1}}{2} \widehat{\boldsymbol{p}}_n = \widehat{\boldsymbol{\pi}}^T \frac{\widetilde{M}_q^{-1}}{2} \widehat{\boldsymbol{\pi}} \tag{1.51}$$

$$\widehat{\boldsymbol{\pi}} = \widetilde{T}^T \widehat{\boldsymbol{p}}_n \tag{1.52}$$

$$\widetilde{M}_q = \widetilde{T}^T \widetilde{M}_n \widetilde{T} \tag{1.53}$$

$$\left[\widetilde{T}\right]_{l,k} = \frac{\partial[\boldsymbol{r}_n]_l}{\partial[\boldsymbol{q}]_k} \tag{1.54}$$

where, within the coordinate eigenstates basis set, the mass tensor $\widetilde{M}_q$ and the transformation matrix $\widetilde{T}$ are identical to the classical corresponding expressions and

$$[\widehat{\boldsymbol{\pi}}]_k = -i\hbar \sum_l \frac{\partial[\boldsymbol{r}_n]_l}{\partial[\boldsymbol{q}]_k} \frac{\partial}{\partial[\boldsymbol{r}_n]_l} = -i\hbar \frac{\partial}{\partial[\boldsymbol{q}]_k} \tag{1.55}$$

provides the components of the new conjugate momenta. Moreover, from $\widehat{\boldsymbol{\pi}}^T = \widehat{\boldsymbol{p}}_n^T \widetilde{T}$ we have

$$\begin{aligned} \left[\widehat{\boldsymbol{\pi}}^T\right]_k &= -i\hbar \sum_l \left(\frac{\partial[\widetilde{T}]_{l,k}}{\partial[\boldsymbol{r}_n]_l} + \frac{\partial[\boldsymbol{r}_n]_l}{\partial[\boldsymbol{q}]_k} \frac{\partial}{\partial[\boldsymbol{r}_n]_l} \right) \\ &= -i\hbar \sum_l \frac{\partial[\widetilde{T}]_{l,k}}{\partial[\boldsymbol{r}_n]_l} - i\hbar \frac{\partial}{\partial[\boldsymbol{q}]_k} \end{aligned} \tag{1.56}$$

clearly showing that in general $[\widehat{\boldsymbol{\pi}}]_k \neq \left[\widehat{\boldsymbol{\pi}}^T\right]_k$ and hence realizing that $\left[\widehat{\boldsymbol{\pi}}^T\right]_k$ is the adjoint of $[\widehat{\boldsymbol{\pi}}]_k$ it follows that the conjugate momenta operators $\widehat{\boldsymbol{\pi}}$ are not always Hermitian operators thus not necessarily corresponding to physical observables. Note that in order to always obtain Hermitian momenta operators we should use

$\left([\widehat{\boldsymbol{\pi}}]_k + [\widehat{\boldsymbol{\pi}}^T]_k\right)/2$ which however would rather complicate the kinetic energy expression; for this reason we keep the definition of the conjugate momenta operators given in Eqs. 1.55 and 1.56. The use of the generalized coordinates $\boldsymbol{q}$ is of particular interest when we can identify a semiclassical subset of such coordinates, i.e., $\boldsymbol{q} = \{\boldsymbol{\xi}, \boldsymbol{\beta}\}$ where $\boldsymbol{\xi}$ are semiclassical nuclear coordinates (usually present when dealing with not too low temperature conditions, in particular for soft condensed phase systems) and $\boldsymbol{\beta}$ are quantum nuclear coordinates. It is worth to note that with semiclassical coordinates we mean nuclear coordinates which behave as virtually classical coordinates, i.e., in any observed system quantum state of interest they are always well localized, and with quantum coordinates we mean the other nuclear coordinates which cannot be considered within such a classical limit. For sake of simplicity we assume that we can use the same $\boldsymbol{\xi}, \boldsymbol{\beta}$ coordinates for all the electron-spin eigenstates of interest (i.e., no mixing of semiclassical and quantum coordinate subspaces), always including the ground eigenstate. In fact, by using the generalized coordinates we can identify for each electron-spin eigenstate the $\boldsymbol{\beta} = \boldsymbol{\beta}_0(\boldsymbol{\xi})$ surface in the nuclear configurational space corresponding to the *molecular* structures (i.e., in general atomic and/or molecular aggregates) where for any possible fixed $\boldsymbol{\xi}$ position the quantum coordinates $\boldsymbol{\beta}$ are relaxed to the minimum electron-spin energy configuration. Note that typically (i.e., rigid/semirigid or usual flexible *molecules*) such $\boldsymbol{\beta}_0(\boldsymbol{\xi})$ surfaces are characterized by tiny variations being all very close to the reference surface $\boldsymbol{\beta} = \boldsymbol{0}$ (for a single eigenstate even possibly coinciding) which hence corresponds for all the electron-spin eigenstates of interest to an approximately relaxed $\boldsymbol{\beta}$ configuration at each $\boldsymbol{\xi}$ position (i.e., approximately equivalent to an ideal physical constraint surface [2] where the semiclassical motions occur). In general, it is possible to use the center of mass (laboratory reference frame) Cartesian coordinates and the Eulerian angles for the *molecular* roto-translational (semiclassical) degrees of freedom and, when dealing with rigid or semirigid harmonic-like *molecules*, we can express the *molecular* (quantum and semiclassical) internal coordinates by means of the simple and very convenient normal mode coordinates as obtained, for the gas-phase *molecular* ground electron-spin eigenstate, within the (orthonormal) *molecular* barycentric reference frame. The use of such quasi-harmonic $\{\boldsymbol{\xi}, \boldsymbol{\beta}\}$ coordinates can be easily extended to flexible *molecules* involving conformations corresponding to harmonic-like wells, by treating each conformation as a semirigid harmonic-like molecular species.

We assume that for any reasonable definition of the $\boldsymbol{\xi}$ coordinates a single $\boldsymbol{\beta}_0$ surface for each electron-spin eigenstate can be used, thus obtaining for any $\boldsymbol{\xi}$ position a unique assignment of the narrow beta coordinates region to be considered (i.e., the accessible beta coordinates region around $\boldsymbol{\beta}_0$). In fact, we can define the $\{\boldsymbol{\xi}, \boldsymbol{\beta}\}$ coordinates according to the chosen *molecular* assignment, considering only the subspace corresponding to a given chemical bonding condition, i.e., constraining each *molecule* to include a given set of nuclei within a specific topology and chemical connectivity. Therefore, it is avoided any chemical bonding variation or identical nuclei permutation, except for the permutations and possibly chemical bond modifications provided by the changes of the semiclassical coordinates with $\boldsymbol{\beta}$ identical or close to $\boldsymbol{\beta}_0$ (i.e., without altering the nuclei assignment and basic chemical bonding structure of each *molecule*). For each chemical bonding condition considered (providing the system chemistry and/or the *molecular* nuclei topology) we can then define the proper $\{\boldsymbol{\xi}, \boldsymbol{\beta}\}$ coordinates and corresponding domain to be used. In practice, once assigned for each *molecule* the corresponding nuclei and related bonding structure, we can use the *molecular* reference geometries (for each *molecule*, typically, the gas-phase electron-spin ground eigenstate optimized geometry) to define the Eulerian angles of each *molecule* providing the rotational orientation of the *molecular* barycentric reference frame with respect to the fixed laboratory reference frame [2]. In such a way we can separate the *molecular* (semiclassical) roto-translational degrees of freedom (the center of mass position $\boldsymbol{R}_G$ and the Eulerian angles θ, ϕ, ψ) from the *molecular* internal Cartesian coordinates given by the nuclei positional deviations from the reference geometry within the *molecular* barycentric reference frame. Note that 6 of such nuclei positional deviations can be expressed as linear combinations of the others as a consequence of the 6 roto-translational coordinates defining the *molecular* barycentric reference frame [2, 3], and thus for a *molecule* including N nuclei we have $3N - 6$ internal Cartesian coordinates (for a rigid linear molecule $3N - 5$). In order to define for each *molecule* the corresponding semiclassical and quantum internal coordinates, once identified a reasonable set of *molecular* internal semiclassical degrees of freedom $\boldsymbol{x}$ (fully defined by the internal Cartesian coordinates, e.g., the dihedral angles), we need to identify the *molecular* reference surface as obtained by minimizing (within the internal Cartesian space and hence at invariant roto-translational coordinates) at each fixed $\boldsymbol{x}$ configuration the gas-phase *molecular* electron-spin ground eigenstate energy [4]. At each position over the *molecular* reference surface we can then define a proper set of local *molecular* quantum coordinates, for the nth *molecule* $\boldsymbol{\beta}_n$, expressing the shifts from such a surface (i.e., $\boldsymbol{\beta}_n = \boldsymbol{0}$ on the nth *molecule* reference surface). Therefore, the *molecular* semiclassical internal coordinates, for the nth *molecule* $\boldsymbol{\xi}_{n,in}$, are defined as the complementary coordinates identical to $\boldsymbol{x}$ at least on the *molecular* reference surface (i.e., at $\boldsymbol{\beta}_n = \boldsymbol{0}$). The complete set of the $\boldsymbol{\xi}, \boldsymbol{\beta}$ coordinates is then obtained collecting all the *molecular* semiclassical and quantum coordinates (i.e., $\boldsymbol{R}_{G,n}, \theta_n, \phi_n, \psi_n, \boldsymbol{\xi}_{n,in}, \boldsymbol{\beta}_n$). The combined effects of the chosen definition of the quantum coordinates $\boldsymbol{\beta}$, the electron-spin eigenstate considered and the intermolecular interactions can clearly cause that $\boldsymbol{\beta}_{0,l}(\boldsymbol{\xi}) \neq \boldsymbol{0}$ (i.e., the lth electron-spin eigenstate energy minima in the $\boldsymbol{\beta}$ subspace at $\boldsymbol{\xi}$ fixed do not belong to the reference surface). However, as previously mentioned, such deviations are typically almost negligible, thus ensuring for the obtained $\boldsymbol{\xi}$ coordinates the proper semiclassical behavior (i.e., the semiclassical motions can be conceived as being virtually confined within the reference surface). Note that in chemical systems with strong hydrogen bonding network, according to the approximation level required, sometimes the proper *molecules* must be defined by tightly bound molecular aggregates.

It is worth to remark that for symmetric or antisymmetric particles (i.e., the particles involved in atomic–molecular systems) the Hamiltonian eigenstates must be symmetric/antisymmetric for identical particles permutations as a consequence of the commuting Hamiltonian and permutation operators, thus requiring in principle to consider all the possible identical nuclei permutations. However, it is easy to realize that for each electron-spin eigenstate the permutations due to the quantum coordinates provide fully equivalent (i.e., physically indistinguishable) *molecular* structures, separated by high energy barriers with hence the $\boldsymbol{\beta}$ coordinates significantly distributed only within a very limited range around each of the equivalent minimum energy positions due to such permutations. Therefore, any different chemical bonding condition due to identical nuclei permutations provides physically equivalent (approximated) eigenstates properly reproducing the physical properties of the exact Hamiltonian eigenstates, thus allowing the use of the $\{\boldsymbol{\xi}, \boldsymbol{\beta}\}$ coordinates and domain corresponding to a single chemical bonding structure. Such an approximation, although equivalent to make physically distinguishable all the nuclei and hence resulting in non-symmetric/antisymmetric approximated Hamiltonian eigenstates for nuclear permutations, provides an accurate description of the exact eigenstates as these almost vanish outside the allowed coordinate regions. The special case of multiple physically distinguishable minima in the $\boldsymbol{\beta}$ subspace at identical $\boldsymbol{\xi}$ position, being very uncommon, will not be addressed in the present derivation.

Analytical mechanics [2] teaches us that the generalized coordinates $\{\boldsymbol{\xi}, \boldsymbol{\beta}\}$ can be chosen in such a way that over the reference $\boldsymbol{\beta} = \mathbf{0}$ surface the mass tensor $\widetilde{M}_q$ becomes a block diagonal matrix

$$\widetilde{M}_q = \begin{bmatrix} \widetilde{M}_{\xi,\xi} & \widetilde{0} \\ \widetilde{0} & \widetilde{M}_{\beta,\beta} \end{bmatrix} \tag{1.57}$$

providing

$$\widehat{K}_n(\boldsymbol{\xi}, \mathbf{0}) = \widehat{K}_\xi(\boldsymbol{\xi}, \mathbf{0}) + \widehat{K}_\beta(\boldsymbol{\xi}, \mathbf{0}) \tag{1.58}$$

$$\widehat{K}_\xi(\boldsymbol{\xi}, \mathbf{0}) = \widehat{\boldsymbol{\pi}}_\xi^T \frac{\widetilde{M}_{\xi,\xi}^{-1}}{2} \widehat{\boldsymbol{\pi}}_\xi \tag{1.59}$$

$$\widehat{K}_\beta(\boldsymbol{\xi}, \mathbf{0}) = \widehat{\boldsymbol{\pi}}_\beta^T \frac{\widetilde{M}_{\beta,\beta}^{-1}}{2} \widehat{\boldsymbol{\pi}}_\beta \tag{1.60}$$

with $\widehat{\boldsymbol{\pi}}_\xi, \widehat{\boldsymbol{\pi}}_\beta$ the conjugate momenta operators for the $\boldsymbol{\xi}$ and $\boldsymbol{\beta}$ coordinates and $\widetilde{M}_{\xi,\xi}, \widetilde{M}_{\beta,\beta}$ the corresponding mass tensor blocks as obtained at $\boldsymbol{\xi}$ and $\boldsymbol{\beta} = \mathbf{0}$ (note that $\widetilde{M}_{\xi,\xi}$ and $\widetilde{M}_{\beta,\beta}$ are also block diagonal tensors with each of their blocks corresponding to a single *molecule*). Therefore, considering that for any accessible configuration (i.e., configurations with a non-negligible probability) and lth electron-spin eigenstate we have tiny $\boldsymbol{\beta}$ deviations from $\boldsymbol{\beta}_{0,l}(\boldsymbol{\xi})$ and hence from $\boldsymbol{\beta} = \mathbf{0}$, we can use $\boldsymbol{\beta} \approx \boldsymbol{\beta}_{0,l}(\boldsymbol{\xi}) \approx \mathbf{0}$ when expressing the kinetic energy operator

$$\widehat{K}_n(\boldsymbol{\xi}, \boldsymbol{\beta}) \cong \widehat{K}_\xi(\boldsymbol{\xi}, \mathbf{0}) + \widehat{K}_\beta(\boldsymbol{\xi}, \mathbf{0}) \tag{1.61}$$

Equation 1.61 states that when semiclassical coordinates are present the proper definition of the generalized coordinates to be used (i.e., the semiclassical $\boldsymbol{\xi}$ and quantum $\boldsymbol{\beta}$ coordinates) allows us to treat the nuclear kinetic energy operator as the sum of two independent terms. The unusual case of flexible *molecules* involving $\boldsymbol{\xi}$

transitions (i.e., conformational transitions) not allowing the use of Eq. 1.61 can be still treated within the outlined framework when conceiving such conformations as different *molecular* species requiring different *molecular* generalized coordinates to fulfill Eq. 1.61. Therefore, for sake of simplicity and without loss of generality, we will not explicitly treat such a case.

When using Eq. 1.61 into Eq. 1.48 we then obtain (for all the Hamiltonian eigenstates involving electron-spin eigenstates sharing the same $\boldsymbol{\xi}, \boldsymbol{\beta}$ coordinates)

$$\left[\widehat{K}_{\xi}(\boldsymbol{\xi}, \mathbf{0}) + \mathcal{U}_{es,l}(\boldsymbol{\xi}, \boldsymbol{\beta}_{0,l}(\boldsymbol{\xi})) + \widehat{H}_{v,l}(\boldsymbol{\xi}, \boldsymbol{\beta})\right] w_{l,k}(\boldsymbol{\xi}, \boldsymbol{\beta}) \cong \mathcal{U}_{l,k} w_{l,k}(\boldsymbol{\xi}, \boldsymbol{\beta}) \quad (1.62)$$

$$\widehat{H}_{v,l}(\boldsymbol{\xi}, \boldsymbol{\beta}) = \widehat{K}_{\beta}(\boldsymbol{\xi}, \mathbf{0}) + \mathcal{U}_{es,l}(\boldsymbol{\xi}, \boldsymbol{\beta}) - \mathcal{U}_{es,l}(\boldsymbol{\xi}, \boldsymbol{\beta}_{0,l}(\boldsymbol{\xi})) \quad (1.63)$$

In Eq. 1.62 the vibrational Hamiltonian operator $\widehat{H}_{v,l}$ which is independent of the $\boldsymbol{\xi}$ conjugate momenta operators commutes with the $\boldsymbol{\xi}$ coordinates and hence its eigenstates, similarly to the electron-spin Hamiltonian eigenstates, can be obtained by $\phi_{v,l}(\boldsymbol{\xi}', \boldsymbol{\beta})\delta(\boldsymbol{\xi} - \boldsymbol{\xi}')$ providing

$$\widehat{H}_{v,l}(\boldsymbol{\xi}, \boldsymbol{\beta})\, \phi_{v,l}(\boldsymbol{\xi}', \boldsymbol{\beta})\delta(\boldsymbol{\xi} - \boldsymbol{\xi}') = \mathcal{U}_{v,l}(\boldsymbol{\xi}')\, \phi_{v,l}(\boldsymbol{\xi}', \boldsymbol{\beta})\delta(\boldsymbol{\xi} - \boldsymbol{\xi}') \quad (1.64)$$

once $\phi_{v,l}$ is defined by

$$\widehat{H}_{v,l}(\boldsymbol{\xi}', \boldsymbol{\beta})\, \phi_{v,l}(\boldsymbol{\xi}', \boldsymbol{\beta}) = \mathcal{U}_{v,l}(\boldsymbol{\xi}')\, \phi_{v,l}(\boldsymbol{\xi}', \boldsymbol{\beta}) \quad (1.65)$$

and hence it corresponds to the vibrational Hamiltonian eigenstate with eigenvalue $\mathcal{U}_{v,l}$, as obtained considering the semiclassical coordinates fixed at the $\boldsymbol{\xi}'$ positions (i.e., the semiclassical coordinates are regarded as physical parameters). Each l, k nuclear eigenstate $w_{l,k}$ (necessarily a finite length quantum state) can then be defined as a linear combination of the $\widehat{H}_{v,l}$ eigenstates (i.e., using the basis set $\phi_{v,l}(\boldsymbol{\xi}', \boldsymbol{\beta})\, \delta(\boldsymbol{\xi} - \boldsymbol{\xi}')$) furnishing

$$\begin{aligned} w_{l,k}(\boldsymbol{\xi}, \boldsymbol{\beta}) &= \sum_{k'} \int a_{l,k,k'}(\boldsymbol{\xi}')\, \phi_{v,l,k'}(\boldsymbol{\xi}', \boldsymbol{\beta})\, \delta(\boldsymbol{\xi} - \boldsymbol{\xi}')\, d\boldsymbol{\xi}' \\ &= \sum_{k'} a_{l,k,k'}(\boldsymbol{\xi})\, \phi_{v,l,k'}(\boldsymbol{\xi}, \boldsymbol{\beta}) \end{aligned} \quad (1.66)$$

The fact that the $\boldsymbol{\xi}$ coordinates can be always considered as semiclassical degrees of freedom with (classical) conjugate momenta $\boldsymbol{\pi}_{\xi}$ necessarily implies that they must be treated via a wave packet and hence (expressing the $\boldsymbol{\xi}, \boldsymbol{\pi}_{\xi}$ phase space position corresponding to the $w_{l,k}$ eigenstate as $\boldsymbol{\tau} = \boldsymbol{\xi}_{\tau}, \boldsymbol{\pi}_{\tau}$) we have

$$a_{l,k,k'}(\boldsymbol{\xi}) \cong A_{\boldsymbol{\tau}}(\boldsymbol{\xi})\, e^{iS_{\tau}(\boldsymbol{\xi})/\hbar}\, a'_{l,k,k'} \quad (1.67)$$

where $a'_{l,k,k'}$ are now constant coefficients and $A_{\boldsymbol{\tau}}(\boldsymbol{\xi})\, e^{iS_{\tau}(\boldsymbol{\xi})/\hbar}$ is the wave packet providing the wave function of the $\xi_j = [\boldsymbol{\xi}]_j$ semiclassical coordinates involved in the nuclear eigenstate with $A_\tau(\boldsymbol{\xi})$ and $S_\tau(\boldsymbol{\xi})$ two real functions with the former, providing the amplitude of the wave packet, highly peaked and virtually confined within a tiny $\boldsymbol{\xi}$ volume around $\boldsymbol{\xi}_\tau$ (i.e., the mean $\boldsymbol{\xi}$ position). Note that for each $\boldsymbol{\xi}$ jth component the corresponding semiclassical conjugate momentum operator is given by [1] $\partial S_\tau/\partial \xi_j$ and hence the wave packet expectation value of the semiclassical conjugate momenta jth component $\langle \partial S_\tau/\partial \xi_j \rangle$ can be expressed by

$$\langle \partial S_\tau/\partial \xi_j \rangle \cong \left(\frac{\partial S_\tau}{\partial \xi_j} \right)_{\xi=\xi_\tau} = [\boldsymbol{\pi}_\tau]_j \tag{1.68}$$

explicitly showing that the wave packet corresponds to a quantum state confined within a tiny $\boldsymbol{\xi}, \boldsymbol{\pi}_\xi$ phase space volume centered at $\boldsymbol{\xi}_\tau, \boldsymbol{\pi}_\tau$ thus providing the best quantum approximation to a classical state.

Equation 1.66 can then be rewritten as

$$w_{l,k}(\boldsymbol{\xi}, \boldsymbol{\beta}) \cong A_\tau \, e^{iS_\tau/\hbar} \sum_{k'} a'_{l,k,k'} \, \phi_{v,l,k'}(\boldsymbol{\xi}, \boldsymbol{\beta}) \tag{1.69}$$

and hence when inserted into Eq. 1.62, multiplying on the left by $e^{-iS_\tau/\hbar}$, provides

$$\begin{aligned} & e^{-iS_\tau/\hbar} \widehat{K}_\xi(\boldsymbol{\xi}, \mathbf{0}) e^{iS_\tau/\hbar} A_\tau \sum_{k'} a'_{l,k,k'} \, \phi_{v,l,k'}(\boldsymbol{\xi}, \boldsymbol{\beta}) \\ & + A_\tau \sum_{k'} \left[\mathcal{U}_{v,l,k'}(\boldsymbol{\xi}) + \mathcal{U}_{es,l}(\boldsymbol{\xi}, \boldsymbol{\beta}_{0,l}(\boldsymbol{\xi})) \right] a'_{l,k,k'} \, \phi_{v,l,k'}(\boldsymbol{\xi}, \boldsymbol{\beta}) \\ \cong \; & \mathcal{U}_{l,k} \, A_\tau \sum_{k'} a'_{l,k,k'} \, \phi_{v,l,k'}(\boldsymbol{\xi}, \boldsymbol{\beta}) \end{aligned} \tag{1.70}$$

The last equation can be considerably simplified by employing

$$\begin{aligned} e^{-iS_\tau/\hbar} \widehat{K}_\xi(\boldsymbol{\xi}, \mathbf{0}) e^{iS_\tau/\hbar} &= e^{-iS_\tau/\hbar} \widehat{\boldsymbol{\pi}}_\xi^T e^{iS_\tau/\hbar} e^{-iS_\tau/\hbar} \frac{\widetilde{M}_{\xi,\xi}^{-1}}{2} e^{iS_\tau/\hbar} e^{-iS_\tau/\hbar} \widehat{\boldsymbol{\pi}}_\xi e^{iS_\tau/\hbar} \\ &= e^{-iS_\tau/\hbar} \widehat{\boldsymbol{\pi}}_\xi^T e^{iS_\tau/\hbar} \frac{\widetilde{M}_{\xi,\xi}^{-1}}{2} e^{-iS_\tau/\hbar} \widehat{\boldsymbol{\pi}}_\xi e^{iS_\tau/\hbar} \end{aligned} \tag{1.71}$$

and thus by using Eqs. 1.55 and 1.56 we can write for each jth component of $\widehat{\boldsymbol{\pi}}_\xi$ and $\widehat{\boldsymbol{\pi}}_\xi^T$

$$\begin{aligned} e^{-iS_\tau/\hbar} \left[\widehat{\boldsymbol{\pi}}_\xi \right]_j e^{iS_\tau/\hbar} &= e^{-iS_\tau/\hbar} \left(-i\hbar \frac{\partial}{\partial [\boldsymbol{\xi}]_j} \right) e^{iS_\tau/\hbar} \\ &= \frac{\partial S_\tau}{\partial [\boldsymbol{\xi}]_j} - i\hbar \frac{\partial}{\partial [\boldsymbol{\xi}]_j} \cong \frac{\partial S_\tau}{\partial [\boldsymbol{\xi}]_j} \end{aligned} \tag{1.72}$$

$$
\begin{aligned}
e^{-iS_\tau/\hbar}\left[\widehat{\boldsymbol{\pi}}_\xi^T\right]_j e^{iS_\tau/\hbar} &= e^{-iS_\tau/\hbar}\left(-i\hbar\sum_g \frac{\partial[\widetilde{T}]_{g,j}}{\partial[\boldsymbol{r}_n]_g} - i\hbar\frac{\partial}{\partial[\boldsymbol{\xi}]_j}\right)e^{iS_\tau/\hbar} \\
&= \frac{\partial S_\tau}{\partial[\boldsymbol{\xi}]_j} - i\hbar\sum_g \frac{\partial[\widetilde{T}]_{g,j}}{\partial[\boldsymbol{r}_n]_g} - i\hbar\frac{\partial}{\partial[\boldsymbol{\xi}]_j} \cong \frac{\partial S_\tau}{\partial[\boldsymbol{\xi}]_j}
\end{aligned} \tag{1.73}
$$

where the index g is running over all the $\boldsymbol{r}_n$ components and we considered as negligible all the terms proportional to $\hbar$ compared to $\partial S_\tau/\partial[\boldsymbol{\xi}]_j$. When using these last equations in Eq. 1.70 (once multiplying on the left the two equation sides by $e^{-iS_\tau/\hbar}$) and considering that the wave packet constrains the semiclassical coordinates to be virtually fixed at $\boldsymbol{\xi} = \boldsymbol{\xi}_\tau$, we obtain

$$
\begin{aligned}
& K_\xi(\boldsymbol{\xi}_\tau, \mathbf{0}, \boldsymbol{\pi}_\tau)\sum_{k'} a'_{l,k,k'}\ \phi_{v,l,k'}(\boldsymbol{\xi}_\tau, \boldsymbol{\beta}) \\
& + \sum_{k'}\left[\mathcal{U}_{v,l,k'}(\boldsymbol{\xi}_\tau) + \mathcal{U}_{es,l}(\boldsymbol{\xi}_\tau, \boldsymbol{\beta}_{0,l}(\boldsymbol{\xi}))\right]\ a'_{l,k,k'}\ \phi_{v,l,k'}(\boldsymbol{\xi}_\tau, \boldsymbol{\beta}) \\
\cong\ & \mathcal{U}_{l,k}\ \sum_{k'} a'_{l,k,k'}\ \phi_{v,l,k'}(\boldsymbol{\xi}_\tau, \boldsymbol{\beta})
\end{aligned} \tag{1.74}
$$

with

$$
K_\xi(\boldsymbol{\xi}_\tau, \mathbf{0}, \boldsymbol{\pi}_\tau) = \boldsymbol{\pi}_\tau^T \frac{\widetilde{M}_{\xi,\xi}^{-1}}{2}\boldsymbol{\pi}_\tau \tag{1.75}
$$

$$
\left[\boldsymbol{\pi}_\tau^T\right]_j = [\boldsymbol{\pi}_\tau]_j = \left(\frac{\partial S_\tau}{\partial \xi_j}\right)_{\xi=\xi_\tau} \tag{1.76}
$$

and clearly $\widetilde{M}_{\xi,\xi}^{-1}$ as obtained at $\boldsymbol{\xi}_\tau$ and $\boldsymbol{\beta} = \mathbf{0}$. Equation 1.74 shows that within the approximations used, the effect of the semiclassical coordinates wave packet $A_\tau\ e^{iS_\tau/\hbar}$ is to provide the $\boldsymbol{\xi}, \widehat{\boldsymbol{\pi}}_\xi \rightarrow \boldsymbol{\xi}_\tau, \boldsymbol{\pi}_\tau$ transformation and, being fulfilled for any accessible $\boldsymbol{\beta}$, implies

$$
\begin{aligned}
& \left[K_\xi(\boldsymbol{\xi}_\tau, \mathbf{0}, \boldsymbol{\pi}_\tau) + \mathcal{U}_{v,l,k'}(\boldsymbol{\xi}_\tau) + \mathcal{U}_{es,l}(\boldsymbol{\xi}_\tau, \boldsymbol{\beta}_{0,l}(\boldsymbol{\xi}))\right]\phi_{v,l,k'}(\boldsymbol{\xi}_\tau, \boldsymbol{\beta}) \\
\cong\ & \mathcal{U}_{l,k}\ \phi_{v,l,k'}(\boldsymbol{\xi}_\tau, \boldsymbol{\beta})
\end{aligned} \tag{1.77}
$$

and hence

$$
\mathcal{U}_{l,k} \cong K_\xi(\boldsymbol{\xi}_\tau, \mathbf{0}, \boldsymbol{\pi}_\tau) + \mathcal{U}_{v,l,k'}(\boldsymbol{\xi}_\tau) + \mathcal{U}_{es,l}(\boldsymbol{\xi}_\tau, \boldsymbol{\beta}_{0,l}(\boldsymbol{\xi})) \tag{1.78}
$$

Therefore, indicating the wave packet via $\phi_\tau(\boldsymbol{\xi})$ (i.e., $\phi_\tau(\boldsymbol{\xi}) = A_\tau(\boldsymbol{\xi})\ e^{iS_\tau(\boldsymbol{\xi})/\hbar}$), we have

$$
\begin{aligned}
& \Phi_{es,l}(\boldsymbol{\xi}, \boldsymbol{\beta}, \boldsymbol{r}_e, \boldsymbol{\gamma}_n, \boldsymbol{\gamma}_e)\ w_{l,k}(\boldsymbol{\xi}, \boldsymbol{\beta}) \cong \Phi_{es,l}(\boldsymbol{\xi}, \boldsymbol{\beta}, \boldsymbol{r}_e, \boldsymbol{\gamma}_n, \boldsymbol{\gamma}_e)\ \phi_{v,l,k'}(\boldsymbol{\xi}, \boldsymbol{\beta})\ \phi_\tau(\boldsymbol{\xi}) \\
\cong\ & \Phi_{es,l}(\boldsymbol{\xi}_\tau, \boldsymbol{\beta}, \boldsymbol{r}_e, \boldsymbol{\gamma}_n, \boldsymbol{\gamma}_e)\ \phi_{v,l,k'}(\boldsymbol{\xi}_\tau, \boldsymbol{\beta})\ \phi_\tau(\boldsymbol{\xi})
\end{aligned} \tag{1.79}
$$

Equation 1.79 clearly indicates that within the approximation of semiclassical $\boldsymbol{\xi}$ coordinates we can express the Born–Oppenheimer Hamiltonian eigenstates $\Phi_{l,k}$, with eigenvalues $\mathcal{U}_{l,k}$, by combining the proper $\boldsymbol{\xi}$ coordinates wave packet with the lth electron-spin eigenstate $\Phi_{es,l}$ and any of the corresponding vibrational eigenstates $\phi_{v,l,k'}$

$$\begin{aligned}\Phi_{l,\boldsymbol{\tau},k'}(\boldsymbol{\xi},\boldsymbol{\beta},\boldsymbol{r}_e,\boldsymbol{\gamma}_n,\boldsymbol{\gamma}_e) &\cong \Phi_{es,l}(\boldsymbol{\xi},\boldsymbol{\beta},\boldsymbol{r}_e,\boldsymbol{\gamma}_n,\boldsymbol{\gamma}_e)\,\phi_{v,l,k'}(\boldsymbol{\xi},\boldsymbol{\beta})\,\phi_{\boldsymbol{\tau}}(\boldsymbol{\xi})\\ &\cong \Phi_{es,l}(\boldsymbol{\xi}_{\boldsymbol{\tau}},\boldsymbol{\beta},\boldsymbol{r}_e,\boldsymbol{\gamma}_n,\boldsymbol{\gamma}_e)\,\phi_{v,l,k'}(\boldsymbol{\xi}_{\boldsymbol{\tau}},\boldsymbol{\beta})\,\phi_{\boldsymbol{\tau}}(\boldsymbol{\xi}) \qquad (1.80)\end{aligned}$$

where the indices l, k' and the phase space position $\boldsymbol{\tau}$ fully identify the Hamiltonian eigenstate $\Phi_{l,\boldsymbol{\tau},k'}$ and eigenvalue $\mathcal{U}_{l,k'}(\boldsymbol{\tau})$ (i.e., we substitute the index k with a combination of $\boldsymbol{\tau}$ and k').

Finally, realizing that for any phase space position wave packet involved into the Hamiltonian eigenstates the corresponding tiny phase space relevant volume (i.e., the phase space volume corresponding to $\boldsymbol{\xi}$, $\partial S_{\boldsymbol{\tau}}/\partial\boldsymbol{\xi}$ values with a non-negligible probability) can be virtually considered a true phase space differential volume, we can safely use the substitution $\boldsymbol{\xi}_{\boldsymbol{\tau}}, \boldsymbol{\pi}_{\boldsymbol{\tau}} \rightarrow \boldsymbol{\xi}, \boldsymbol{\pi}_{\xi}$ (i.e., $\phi_{\boldsymbol{\tau}} = \phi_{\boldsymbol{\xi},\boldsymbol{\pi}_\xi}$) to express the Hamiltonian eigenstates (using now the index k to identify the vibrational eigenstates) via

$$\Phi_{l,\boldsymbol{\xi},\boldsymbol{\pi}_\xi,k}(\boldsymbol{\xi}',\boldsymbol{\beta},\boldsymbol{r}_e,\boldsymbol{\gamma}_n,\boldsymbol{\gamma}_e) \cong \Phi_{es,l}(\boldsymbol{\xi}',\boldsymbol{\beta},\boldsymbol{r}_e,\boldsymbol{\gamma}_n,\boldsymbol{\gamma}_e)\,\phi_{v,l,k}(\boldsymbol{\xi}',\boldsymbol{\beta})\,\phi_{\boldsymbol{\xi},\boldsymbol{\pi}_\xi}(\boldsymbol{\xi}') \qquad (1.81)$$

and thus to write

$$\begin{aligned}&\left[K_\xi(\boldsymbol{\xi},\boldsymbol{0},\boldsymbol{\pi}_\xi)+\widehat{K}_\beta(\boldsymbol{\xi},\boldsymbol{0})+\widehat{H}_{es}(\boldsymbol{\xi},\boldsymbol{\beta},\boldsymbol{r}_e,\widehat{\boldsymbol{p}}_e,\widetilde{\boldsymbol{s}}_n,\widetilde{\boldsymbol{s}}_e)\right]\Phi_{es,l}(\boldsymbol{\xi},\boldsymbol{\beta},\boldsymbol{r}_e,\boldsymbol{\gamma}_n,\boldsymbol{\gamma}_e)\,\phi_{v,l,k}(\boldsymbol{\xi},\boldsymbol{\beta})\\ &\cong \mathcal{U}_{l,k}(\boldsymbol{\xi},\boldsymbol{\pi}_\xi)\,\Phi_{es,l}(\boldsymbol{\xi},\boldsymbol{\beta},\boldsymbol{r}_e,\boldsymbol{\gamma}_n,\boldsymbol{\gamma}_e)\,\phi_{v,l,k}(\boldsymbol{\xi},\boldsymbol{\beta}) \qquad (1.82)\end{aligned}$$

with

$$\mathcal{U}_{l,k}(\boldsymbol{\xi},\boldsymbol{\pi}_\xi) \cong K_\xi(\boldsymbol{\xi},\boldsymbol{0},\boldsymbol{\pi}_\xi)+\mathcal{U}_{v,l,k}(\boldsymbol{\xi})+\mathcal{U}_{es,l}(\boldsymbol{\xi},\boldsymbol{\beta}_{0,l}(\boldsymbol{\xi})) \qquad (1.83)$$

$$K_\xi(\boldsymbol{\xi},\boldsymbol{0},\boldsymbol{\pi}_\xi) = \boldsymbol{\pi}_\xi^T\frac{\widetilde{M}_{\xi,\xi}^{-1}}{2}\boldsymbol{\pi}_\xi \qquad (1.84)$$

Note that Eq. 1.81 provides a proper approximation of the Hamiltonian eigenstates for a system with semiclassical coordinates, when considering that the obtained degenerate eigenstates corresponding to identical nuclei permutations due to different $\boldsymbol{\xi}$ configurations within the considered $\boldsymbol{\xi}, \boldsymbol{\beta}$ domain (i.e., without altering the chosen nuclei assignment and basic chemical bonding structure of each *molecule*), must be considered as physically indistinguishable states in order to fulfill the symmetry/antisymmetry property of the exact Hamiltonian eigenstates. Therefore, they should be always counted as a single eigenstate defined by a proper linear combination of them [1].

The Hamiltonian eigenstates $\Phi_{l,\boldsymbol{\xi},\boldsymbol{\pi}_\xi,k}$ are normalized wavefunctions and thus (using normalized electron-spin eigenstates and the Jacobean $J(\boldsymbol{\xi}', \boldsymbol{\beta})$ for the $\boldsymbol{r}_n \to \boldsymbol{\xi}', \boldsymbol{\beta}$ transformation)

$$\begin{aligned} 1 &= \sum_{\gamma_n,\gamma_e} \int |\Phi_{es,l}\, \phi_{v,l,k}\, \phi_{\boldsymbol{\xi},\boldsymbol{\pi}_\xi}|^2 d\boldsymbol{r}_e\, d\boldsymbol{r}_n = \sum_{\gamma_n,\gamma_e} \int |\Phi_{es,l}\, \phi_{v,l,k}\, \phi_{\boldsymbol{\xi},\boldsymbol{\pi}_\xi}|^2 J(\boldsymbol{\xi}', \boldsymbol{\beta})\, d\boldsymbol{r}_e\, d\boldsymbol{\xi}'\, d\boldsymbol{\beta} \\ &\cong \sum_{\gamma_n,\gamma_e} \int |\Phi_{es,l}(\boldsymbol{\xi}, \boldsymbol{\beta}, \boldsymbol{r}_e, \boldsymbol{\gamma}_n, \boldsymbol{\gamma}_e)\, \phi_{v,l,k}(\boldsymbol{\xi}, \boldsymbol{\beta})\, \phi_{\boldsymbol{\xi},\boldsymbol{\pi}_\xi}(\boldsymbol{\xi}')|^2 J(\boldsymbol{\xi}, \boldsymbol{\beta}_{0,l}(\boldsymbol{\xi}))\, d\boldsymbol{r}_e\, d\boldsymbol{\xi}'\, d\boldsymbol{\beta} \\ &= \left(\int |\phi_{\boldsymbol{\xi},\boldsymbol{\pi}_\xi}(\boldsymbol{\xi}')|^2\, d\boldsymbol{\xi}'\right)\left(J(\boldsymbol{\xi}, \boldsymbol{\beta}_{0,l}(\boldsymbol{\xi})) \int |\phi_{v,l,k}(\boldsymbol{\xi}, \boldsymbol{\beta})|^2\, d\boldsymbol{\beta}\right) \end{aligned} \tag{1.85}$$

showing that when using normalized wave packets (i.e., $\int |\phi_{\boldsymbol{\xi},\boldsymbol{\pi}_\xi}(\boldsymbol{\xi}')|^2\, d\boldsymbol{\xi}' = 1$) we necessarily have

$$J(\boldsymbol{\xi}, \boldsymbol{\beta}_{0,l}(\boldsymbol{\xi})) \int |\phi_{v,l,k}(\boldsymbol{\xi}, \boldsymbol{\beta})|^2\, d\boldsymbol{\beta} \cong 1 \tag{1.86}$$

and thus including $\sqrt{J(\boldsymbol{\xi}, \boldsymbol{\beta}_{0,l}(\boldsymbol{\xi}))}$ into the vibrational eigenstates allows us to disregard the Jacobean, i.e., at each semiclassical configuration $\boldsymbol{\xi}$ we obtain normalized vibrational eigenstates.

References

1. Dirac PAM (1958) The principles of quantum mechanics. Clarendon Press, Oxford
2. Gallavotti G (1983) The elements of mechanics. Springer, New York
3. Amadei A, Chillemi G, Ceruso MA, Grottesi A, Di Nola A (2000) Molecular simulations with constrained roto-translational motions: theoretical basis and statistical mechanical consistency. J Chem Phys 112:9–23
4. At each minimization step the displacement within the internal Cartesian space is taken along the force component orthogonal to the $\boldsymbol{x}$ gradients

Chapter 2
Statistical Mechanics: Basic Principles

Abstract In this chapter we derive the foundational principles of statistical mechanics, as obtained by the density operator dynamics within a proper statistical ensemble. We first introduce the canonical ensemble deriving its statistics and corresponding thermodynamics, illustrating the canonical ensemble special case corresponding to the microcanonical ensemble. We then derive the isothermal–isobaric and grand-canonical ensemble statistics and thermodynamics, comparing the different ensembles and discussing the conditions ensuring the same thermodynamics regardless of the ensemble used.

2.1 General Considerations

In Chemistry the typical systems investigated are macroscopic systems, hence involving a huge number of interacting particles: the electrons and nuclei due to a number of atoms or molecules in the order of Avogadro's number $\mathcal{N}^0 = 6.022\ 10^{23}$. Therefore, the enormous number of degrees of freedom and the complexity of the Hamiltonian impede any attempt to reconstruct the detailed dynamics of the system. By considering, instead, that any macroscopic system can be reasonably conceived as a very large collection of equivalent statistically independent subsystems (the elementary systems) embedded into a virtually infinite thermal bath, we can be sure that the use of a statistical treatment is not only necessary to reduce the complexity of the model but it is also the proper approach to be used for describing the macroscopic system thermodynamics and kinetics [1, 2]. Note that each elementary system, although much smaller than the whole system, must be large enough to avoid any significant effect of the matter fluctuations as well as to disregard its interactions with the environment (i.e., the other elementary systems and the thermal bath) in order to achieve identical (fixed) chemical composition and statistical independence. Such interactions can be then considered as providing only tiny and sudden energy exchanges due to molecular collisions (heat exchanges), determining the transition from one eigenenergy to another without significantly modifying the elementary system Hamiltonian eigenstates and eigenvalues. The statistics of each elementary system then provides all the intensive properties of the macroscopic system (necessarily larger than or at

A. Amadei and M. Aschi, *Statistical Mechanics for Chemical Thermodynamics and Kinetics*, https://doi.org/10.1007/978-3-031-77929-9_2

most equal to an elementary system) with the extensive ones simply obtained by linearly scaling over the number of elementary systems (i.e., the thermodynamic behavior).

2.2 The Density Operator and the Canonical Ensemble

In order to obtain a proper statistical description, on the basis of the previous considerations, we need to introduce the canonical ensemble defined in principle by an infinite set of statistically independent systems of identical fixed chemical composition and volume, with each system in thermal equilibrium with a virtually infinite thermal bath (all these thermal baths are equivalent having identical properties and conditions). Note that any proper ensemble system must be larger than or at most equal to the elementary system, with its particles confined within an ideal rigid box fixing the volume. For such an ensemble we can define the Hermitian operator $\widehat{\rho}$ (i.e., the density operator [1, 2]) via

$$\widehat{\rho}(t) = \frac{1}{n} \sum_{l=1}^{n} |c_l(t)\rangle\langle c_l(t)| \tag{2.1}$$

or in matrix notation (using a discrete basis set to express the operator)

$$\widetilde{\rho}(t) = \frac{1}{n} \sum_{l=1}^{n} \boldsymbol{c}_l(t)\boldsymbol{c}_l^{\dagger}(t) \tag{2.2}$$

where $|c_l(t)\rangle$ is the time-dependent ket representing the (normalized) dynamical quantum state of the lth system in the ensemble and in principle $n \to \infty$ ($\boldsymbol{c}_l$ is clearly the column vector expressing the $|c_l\rangle$ ket within the discrete basis set chosen). From Eq. 2.1 we then have that for any quantum state $|\eta\rangle$ (always conceivable as a member of an orthogonal basis set) the expression

$$\langle\eta|\widehat{\rho}|\eta\rangle = \frac{1}{n} \sum_{l=1}^{n} \langle\eta|c_l(t)\rangle\langle c_l(t)|\eta\rangle = \frac{1}{n} \sum_{l=1}^{n} |\langle\eta|c_l(t)\rangle|^2 \tag{2.3}$$

provides the probability of finding a system of the ensemble in state η at time t. From the definition of the density operator it also follows that for any observable expressed by the operator $\widehat{O}$ we have that its expectation value over the ensemble systems (i.e., the ensemble average) is given by

$$\frac{1}{n} \sum_{l} \langle c_l(t)|\widehat{O}|c_l(t)\rangle = \frac{1}{n} \sum_{l} \sum_{j} \sum_{j'} c_{l,j}^{*}(t) c_{l,j'}(t) \langle\eta_j|\widehat{O}|\eta_{j'}\rangle$$

$$
\begin{aligned}
&= \frac{1}{n}\sum_{l}\sum_{j}\sum_{j'} c^*_{l,j}(t)\langle\eta_{j'}|c_l(t)\rangle\langle\eta_j|\widehat{O}|\eta_{j'}\rangle \\
&= \sum_{j'}\frac{1}{n}\sum_{l}\langle\eta_{j'}|c_l(t)\rangle\sum_{j} c^*_{l,j}(t)\langle\eta_j|\widehat{O}|\eta_{j'}\rangle \\
&= \sum_{j'}\frac{1}{n}\sum_{l}\langle\eta_{j'}|c_l(t)\rangle\langle c_l(t)|\widehat{O}|\eta_{j'}\rangle \\
&= \sum_{j'}\langle\eta_{j'}|\Big[\Big(\frac{1}{n}\sum_{l}|c_l(t)\rangle\langle c_l(t)|\Big)\widehat{O}\Big]|\eta_{j'}\rangle \\
&= \sum_{j'}\langle\eta_{j'}|\widehat{\rho}\,\widehat{O}|\eta_{j'}\rangle = Tr\Big\{\widehat{\rho}\,\widehat{O}\Big\} = Tr\Big\{\widehat{O}\,\widehat{\rho}\Big\} \qquad (2.4)
\end{aligned}
$$

where the $|\eta_j\rangle$ kets define the (discrete) basis set used, $c_{l,j}$ are the components of the $\boldsymbol{c}_l$ column vector and $Tr\Big\{\cdots\Big\}$ represents the trace of the argument operator (note that an identical result would be obtained when using a continuous basis set, although in such a case the summations over the basis set kets should be replaced by integrals). Equations 2.3 and 2.4 clearly show that $\widehat{\rho}$ is the operator representing the probability within the ensemble and thus it is the quantum equivalent of the probability density in phase space for classical mechanics.

When considering for each system in the ensemble the corresponding total Hamiltonian operator

$$\widehat{H}_l(t) = \widehat{H}^0 + \widehat{V}_l(t) \qquad (2.5)$$

with $\widehat{H}^0$ the stationary Hamiltonian of the ensemble system when neglecting any interaction with its environment (the unperturbed Hamiltonian operator identical for all the ensemble systems) and $\widehat{V}_l(t)$ the tiny (i.e., almost negligible compared to the system energy) time-dependent perturbation mimicking the interactions involved in the heat exchanges between the lth system and the bath, we can easily obtain the proper equation for the dynamics of the density operator. In fact, using the ket and bra forms of the Schroedinger equation for the generic $|c_l(t)\rangle$ quantum state

$$i\hbar\frac{\partial}{\partial t}|c_l(t)\rangle = \widehat{H}_l(t)|c_l(t)\rangle \qquad (2.6)$$

$$i\hbar\frac{\partial}{\partial t}\langle c_l(t)| = -\langle c_l(t)|\widehat{H}_l(t) \qquad (2.7)$$

we can write from Eq. 2.1

$$i\hbar\frac{\partial\widehat{\rho}}{\partial t} = i\hbar\frac{1}{n}\sum_{l=1}^{n}\frac{\partial|c_l(t)\rangle}{\partial t}\langle c_l(t)| + i\hbar\frac{1}{n}\sum_{l=1}^{n}|c_l(t)\rangle\frac{\partial\langle c_l(t)|}{\partial t}$$

$$= \widehat{H}^0\widehat{\rho}(t) - \widehat{\rho}(t)\widehat{H}^0 + \frac{1}{n}\sum_{l=1}^{n}\Big[\widehat{V}_l(t)|c_l(t)\rangle\langle c_l(t)| - |c_l(t)\rangle\langle c_l(t)|\widehat{V}_l(t)\Big] \quad (2.8)$$

It is very interesting to consider the last equation in the limit of full statistical equilibrium (i.e., stationary density operator) when it is reasonable to assume that the heat exchanges are virtually statistically independent of the instantaneous state of the system, thus providing

$$\frac{1}{n}\sum_{l=1}^{n}\Big[\widehat{V}_l(t)|c_l(t)\rangle\langle c_l(t)| - |c_l(t)\rangle\langle c_l(t)|\widehat{V}_l(t)\Big] \cong \langle\widehat{V}\rangle\,\widehat{\rho} - \widehat{\rho}\,\langle\widehat{V}\rangle \quad (2.9)$$

where

$$\langle\widehat{V}\rangle = \frac{1}{n}\sum_{l=1}^{n}\widehat{V}_l(t) \quad (2.10)$$

Moreover, when considering that at equilibrium the ensemble average of the heat exchanges should vanish, i.e., $\langle\widehat{V}\rangle = 0$, and $\partial\widehat{\rho}/\partial t = 0$ we readily obtain from Eqs. 2.8–2.10

$$i\hbar\frac{\partial\widehat{\rho}}{\partial t} \cong \widehat{H}^0\widehat{\rho}(t) - \widehat{\rho}(t)\widehat{H}^0 = 0 \quad (2.11)$$

clearly indicating that the density operator at equilibrium commutes with the stationary Hamiltonian $\widehat{H}^0$ and thus the density operator should be a function of $\widehat{H}^0$ and in principle of the other operators, if present, defining with $\widehat{H}^0$ a complete set of commuting observables (see the Appendix). In fact, for a system within a fixed volume as provided by an ideal rigid box (a box with reflective walls for the system particles, i.e., elastic collisions), the total linear and angular momenta are not invariant anymore and hence they cannot commute with the invariant $\widehat{H}^0$, thus indicating that at equilibrium the density operator can be expressed as a function of the unperturbed Hamiltonian only: $\widehat{\rho}$ is diagonal (incoherent) when represented within the $\widehat{H}^0$ eigenstate basis set. Therefore, the corresponding density operator eigenvalues (i.e., the equilibrium probabilities of the $\widehat{H}^0$ eigenstates) must be proportional to a function of the unperturbed energy $\mathcal{U}$, i.e., the energy provided by the eigenvalues of $\widehat{H}^0$, hereafter simply the system energy. We can then express the eigenvalues of the equilibrium density operator via $\rho \cong \alpha F(\mathcal{U})$ with α an ensemble-dependent normalization constant and $F(\mathcal{U})$ a real function fully defined by the thermal bath (i.e., identical for all systems when embedded into equivalent baths). Therefore, for any couple of systems A and B each well described by a canonical ensemble with identical thermal baths, we can write

$$\rho_A \cong \alpha_A F(\mathcal{U}_A) \quad (2.12)$$
$$\rho_B \cong \alpha_B F(\mathcal{U}_B) \quad (2.13)$$

and hence when considering these two statistically independent systems together forming the new AB system we have

$$\rho_{AB} \cong \alpha_{AB} F(\mathcal{U}_{AB}) \cong \rho_A \rho_B \tag{2.14}$$

Realizing that within the approximations used we can neglect any interaction between the A and B systems, i.e., $\mathcal{U}_{AB} \cong \mathcal{U}_A + \mathcal{U}_B$, we necessarily have

$$\alpha_{AB} F(\mathcal{U}_A + \mathcal{U}_B) \cong \alpha_A F(\mathcal{U}_A) \alpha_B F(\mathcal{U}_B) \tag{2.15}$$

plainly indicating that

$$\alpha_{AB} \cong \alpha_A \alpha_B \tag{2.16}$$

$$F(\mathcal{U}_A + \mathcal{U}_B) \cong F(\mathcal{U}_A) F(\mathcal{U}_B) \tag{2.17}$$

A simple choice for $F(\mathcal{U})$ fully fulfilling Eq. 2.17 is the exponential function providing

$$\rho \cong \alpha e^{-\beta \mathcal{U}} \tag{2.18}$$

with β a constant determined by the thermal bath and the α normalization constant obtained from

$$\sum_j \rho_j \cong \alpha \sum_j e^{-\beta \mathcal{U}_j} = 1 \tag{2.19}$$

We can then write for any kind of macroscopic system (i.e., larger than or at most equal to the elementary system) the equilibrium probability for the jth (unperturbed) Hamiltonian eigenstate, according to the canonical ensemble statistics

$$\rho_j \cong \frac{e^{-\beta \mathcal{U}_j}}{Q} \tag{2.20}$$

$$Q = \sum_j e^{-\beta \mathcal{U}_j} \tag{2.21}$$

with Q the canonical partition function, fully defined by the system chemical composition and volume as well as by the β constant determined by the thermal bath. From the last equations it readily follows that the equilibrium canonical density operator can be expressed as

$$\widehat{\rho} = \frac{e^{-\beta \widehat{H}^0}}{Q} \tag{2.22}$$

2.3 The Statistical View of Thermodynamics

2.3.1 The Basic Equations

From Eqs. 2.20 and 2.21 we can obtain the system thermodynamics as provided by the canonical ensemble, by realizing that each Hamiltonian eigenvalue is fully determined by the chemical composition (i.e., the amount of each chemical species) and the volume of the system. Note that only when considering the special case of a system subject to an external electro-magnetic field, the geometry of the box as well as the external field become other state variables affecting the Hamiltonian eigenvalues (we disregard the gravitational field as its effect on atomic–molecular particles is typically too weak). In the present derivation, without loss of generality for the results obtained, we do not consider such a special case and hence we disregard the geometry-field effects. We can then express the equilibrium pressure p via

$$
\begin{aligned}
p &= \sum_j \rho_j \left(-\frac{\partial \mathcal{U}_j}{\partial V}\right)_{\boldsymbol{N}} = -\sum_j \frac{e^{-\beta \mathcal{U}_j}}{Q} \left(\frac{\partial \mathcal{U}_j}{\partial V}\right)_{\boldsymbol{N}} \\
&= \frac{1}{\beta}\left(\frac{\partial \ln Q}{\partial V}\right)_{\boldsymbol{N},\beta}
\end{aligned}
\tag{2.23}
$$

where V is the system volume, $\boldsymbol{N}$ is the vector defined by the molecular amounts (in number of molecules) of all the chemical species and $\left(-\partial \mathcal{U}_j/\partial V\right)_{\boldsymbol{N}}$ clearly provides the pressure of the system when considering a single eigenstate with eigenvalue $\mathcal{U}_j(\boldsymbol{N}, V)$. Similarly, we can express the thermodynamic energy (i.e., the equilibrium average energy) U via

$$
\begin{aligned}
U &= \sum_j \rho_j \mathcal{U}_j = \sum_j \frac{e^{-\beta \mathcal{U}_j}}{Q} \mathcal{U}_j \\
&= -\left(\frac{\partial \ln Q}{\partial \beta}\right)_{\boldsymbol{N},V}
\end{aligned}
\tag{2.24}
$$

and the chemical potential of the lth chemical species μ_l as

$$
\begin{aligned}
\mu_l &= \sum_j \rho_j \left(\frac{\partial \mathcal{U}_j}{\partial N_l}\right)_{\boldsymbol{N}_l,V} = \sum_j \frac{e^{-\beta \mathcal{U}_j}}{Q} \left(\frac{\partial \mathcal{U}_j}{\partial N_l}\right)_{\boldsymbol{N}_l,V} \\
&= -\frac{1}{\beta}\left(\frac{\partial \ln Q}{\partial N_l}\right)_{\boldsymbol{N}_l,V,\beta}
\end{aligned}
\tag{2.25}
$$

where N_l is the number of molecules of the lth species, $\boldsymbol{N}_l$ is the vector of the molecular amounts when removing N_l and by definition $\left(\partial \mathcal{U}_j/\partial N_l\right)_{\boldsymbol{N}_l,V}$ is the corresponding chemical potential when considering a single eigenstate with eigenvalue

$\mathcal{U}_j(N, V)$. Note that we can use the number of molecules as a continuous variable, considering its minimal change of one molecule a differential, as we deal with macroscopic systems and thus $1/\sum_l N_l \to 0$.

Equations 2.23–2.25 clearly indicate that the canonical partition function $Q(N, V, \beta)$ can be related to the Helmholtz free energy $A(N, V, T)$ (with T the absolute temperature) as it follows from the general thermodynamic equations

$$p = -\left(\frac{\partial A}{\partial V}\right)_{N,T} \tag{2.26}$$

$$U = \left(\frac{\partial A/T}{\partial 1/T}\right)_{N,V} \tag{2.27}$$

$$\mu_l = \left(\frac{\partial A}{\partial N_l}\right)_{N_l,V,T} \tag{2.28}$$

2.3.2 The Ideal Gas Canonical Partition Function

To proceed further in establishing the relation between the canonical partition function and the system thermodynamics, we need to unveil the dependence of β on the thermal bath properties. In order to obtain an explicit expression of β in terms of the bath state variables, we can evaluate the partition function for the simplest conceivable canonical system: an ideal gas of N spinless identical particles of mass m, i.e., non interacting particles, confined within an ideal rigid box of volume V and in thermal equilibrium with a bath at temperature T. Each particle of such a system can obviously be considered as a free particle except for its elastic collisions with the reflective walls. Therefore, the Hamiltonian operator of the system is given by the kinetic energy only and its eigenstates can be obtained by combining (within the coordinate representation by multiplying) the eigenstates of the single particles, which in turn can be expressed by means of the free particle linear momentum eigenstates (also eigenstates of the Hamiltonian) with eigenvalues $p_{x,j_x}, p_{y,j_y}, p_{z,j_z}$ as described in Chap. 1. In fact, such free particle eigenstates within the Cartesian coordinate representation, i.e., $\Phi_{j_x,j_y,j_z} = Ce^{\frac{i}{\hbar}(p_{x,jx}x+p_{y,jy}y+p_{z,jz}z)} = \Phi_{x,j_x}\Phi_{y,j_y}\Phi_{z,j_z}$ with C an arbitrary constant (see Chap. 1), are provided by any combination (i.e., multiplication) of the linear momentum single component eigenstates

$$\Phi_{x,j_x} = C_x e^{\frac{i}{\hbar}p_{x,jx}x} \tag{2.29}$$

$$\Phi_{y,j_y} = C_y e^{\frac{i}{\hbar}p_{y,jy}y} \tag{2.30}$$

$$\Phi_{z,j_z} = C_z e^{\frac{i}{\hbar}p_{z,jz}z} \tag{2.31}$$

with clearly $C = C_x C_y C_z$.

Therefore, once considering a cubic box and the laboratory fixed frame with origin at a corner of the cubic box and axes along the cube sides, the Hamiltonian eigenstates of each single particle within the box (no more linear momentum eigenstates) can be expressed again via multiplication of the single coordinate wavefunctions, i.e., $\Phi_{j_x,j_y,j_z} = \Phi_{x,j_x}\Phi_{y,j_y}\Phi_{z,j_z}$, with however each single coordinate wavefunction now given by a linear combination of two free particle single component eigenstates of identical momentum eigenvalue except for its sign

$$\Phi_{x,j_x} = C_x\left\{e^{\frac{i}{\hbar}p_{x,j_x}x} - e^{-\frac{i}{\hbar}p_{x,j_x}x}\right\} = 2iC_x \sin\left(p_{x,j_x}x/\hbar\right) \tag{2.32}$$

$$\Phi_{y,j_y} = C_y\left\{e^{\frac{i}{\hbar}p_{y,j_y}y} - e^{-\frac{i}{\hbar}p_{y,j_y}y}\right\} = 2iC_y \sin\left(p_{y,j_y}y/\hbar\right) \tag{2.33}$$

$$\Phi_{z,j_z} = C_z\left\{e^{\frac{i}{\hbar}p_{z,j_z}z} - e^{-\frac{i}{\hbar}p_{z,j_z}z}\right\} = 2iC_z \sin\left(p_{z,j_z}z/\hbar\right) \tag{2.34}$$

with clearly p_{x,j_x}, p_{y,j_y}, p_{z,j_z} now indicating the norms of the linear momentum component eigenvalues, hence providing

$$\begin{aligned}\Phi_{j_x,j_y,j_z} &= C\left\{e^{\frac{i}{\hbar}p_{x,j_x}x} - e^{-\frac{i}{\hbar}p_{x,j_x}x}\right\}\left\{e^{\frac{i}{\hbar}p_{y,j_y}y} - e^{-\frac{i}{\hbar}p_{y,j_y}y}\right\}\left\{e^{\frac{i}{\hbar}p_{z,j_z}z} - e^{-\frac{i}{\hbar}p_{z,j_z}z}\right\} \\ &= -8iC \sin\left(p_{x,j_x}x/\hbar\right) \sin\left(p_{y,j_y}y/\hbar\right) \sin\left(p_{z,j_z}z/\hbar\right) \end{aligned} \tag{2.35}$$

Note that each of the wavefunctions provided by Eqs. 2.32–2.34 being the over-position of a free particle positive and negative momentum component eigenstates physically corresponds to the free particle motion along a given axis including the effects of the elastic collisions. Such wavefunctions, although no more eigenstates of the momentum components, are each the eigenstate of the corresponding kinetic energy term, e.g., Φ_x is the eigenstate of the $\widehat{K}_x = \widehat{p}_x^2/(2m)$ operator, thus guaranteeing that Eq. 2.35 provides the single particle Hamiltonian eigenstates (for an ideal gas particle the kinetic energy operator, $\widehat{K} = \widehat{K}_x + \widehat{K}_y + \widehat{K}_z$, coincides with the Hamiltonian operator). The particle wavefunction given by Eq. 2.35 is then the proper expression for the particle Hamiltonian eigenstates within the box (outside the box the eigenstate wavefunctions should virtually vanish as the box walls correspond to virtually infinite energy barriers) and hence it should vanish on the box walls, i.e., at $x = 0, L_x$ or $y = 0, L_y$ or $z = 0, L_z$ with L_x, L_y, L_z the lengths of the box sides. Therefore, we necessarily must have

$$p_{x,j_x} = \frac{j_x h}{2L_x} \tag{2.36}$$

$$p_{y,j_y} = \frac{j_y h}{2L_y} \tag{2.37}$$

$$p_{z,j_z} = \frac{j_z h}{2L_z} \tag{2.38}$$

with j_x, j_y, j_z any natural number. Realizing that for any macroscopic system the variation of the norm of each momentum component eigenvalue for the $j \rightarrow j+1$ change is virtually a differential, we obtain (once defining the virtually continuous variables $p_x = j_x h/(2L_x) \geq 0$, $p_y = j_y h/(2L_y) \geq 0$, $p_z = j_z h/(2L_z) \geq 0$)

$$\sum_{j_x=0}^{\infty} e^{-\beta K_{x,jx}} = \sum_{j_x=0}^{\infty} e^{-\beta \frac{p_{x,jx}^2}{2m}} \cong \int_0^{\infty} e^{-\beta(\frac{j_x h}{2L_x})^2/(2m)} \, dj_x$$
$$= \frac{2L_x}{h} \int_0^{\infty} e^{-\beta \frac{p_x^2}{2m}} \, dp_x = \frac{L_x}{h} \left[2\pi m/\beta\right]^{1/2} \tag{2.39}$$

$$\sum_{j_y=0}^{\infty} e^{-\beta K_{y,jy}} = \sum_{j_y=0}^{\infty} e^{-\beta \frac{p_{y,jy}^2}{2m}} \cong \int_0^{\infty} e^{-\beta(\frac{j_y h}{2L_y})^2/(2m)} \, dj_y$$
$$= \frac{2L_y}{h} \int_0^{\infty} e^{-\beta \frac{p_y^2}{2m}} \, dp_y = \frac{L_y}{h} \left[2\pi m/\beta\right]^{1/2} \tag{2.40}$$

$$\sum_{j_z=0}^{\infty} e^{-\beta K_{z,jz}} = \sum_{j_z=0}^{\infty} e^{-\beta \frac{p_{z,jz}^2}{2m}} \cong \int_0^{\infty} e^{-\beta(\frac{j_z h}{2L_z})^2/(2m)} \, dj_z$$
$$= \frac{2L_z}{h} \int_0^{\infty} e^{-\beta \frac{p_z^2}{2m}} \, dp_z = \frac{L_z}{h} \left[2\pi m/\beta\right]^{1/2} \tag{2.41}$$

where K_{x,j_x}, K_{y,j_y}, K_{z,j_z} are the eigenvalues of the $\widehat{K}_x$, $\widehat{K}_y$, $\widehat{K}_z$ operator, respectively. From the last equations we can write the single particle partition function q via

$$q = \Big(\sum_{j_x=0}^{\infty} e^{-\beta K_{x,jx}}\Big)\Big(\sum_{j_y=0}^{\infty} e^{-\beta K_{y,jy}}\Big)\Big(\sum_{j_z=0}^{\infty} e^{-\beta K_{z,jz}}\Big) \cong \frac{V}{h^3}\left[2\pi m/\beta\right]^{3/2} \tag{2.42}$$

(with $V = L_x L_y L_z$ the box volume) and thus the canonical partition function Q of the whole system made of N identical spinless particles in the ideal gas condition

$$Q \cong \frac{q^N}{N!} \cong \frac{V^N}{h^{3N} N!}\left[2\pi m/\beta\right]^{3N/2} \tag{2.43}$$

where $N!$ is introduced to correct the summation over all the combinations of the particle eigenstates (i.e., q^N) from those terms corresponding to permutations of identical particles and hence providing physically non-distinguishable states: for the system considered at not too low temperatures given the almost continuous nature of the Hamiltonian eigenvalues we can safely disregard all the combinations with more than a particle in a given particle eigenstate, thus allowing the factorial as a proper correction. By using Eq. 2.43 into Eq. 2.23 we then obtain

$$p = \frac{1}{\beta}\Big(\frac{\partial \ln Q}{\partial V}\Big)_{N,\beta} = \frac{N}{\beta V} \tag{2.44}$$

and thus when comparing this last result with the (empirical) ideal gas law for a system of N particles at temperature T within the V volume, i.e., $p = N k_B T/V$ with k_B the Boltzmann constant, we can readily write $1/\beta = k_B T$ which must be valid for any possible system (e.g., made of any kind of interacting particles) at temperature T (i.e., embedded within a thermal bath at temperature T).

Finally, it is useful to consider a classical-like description of the ideal gas system with hence the (semiclassical) Hamiltonian eigenstates corresponding to wave packets (see Chap. 1), each confined within a tiny phase space volume (approximating a differential volume) identified by the wave packet mean coordinates and corresponding (semiclassical) momenta

$$\boldsymbol{r}_l = \begin{bmatrix} r_{x,l} \\ r_{y,l} \\ r_{z,l} \end{bmatrix} \quad \boldsymbol{p}_l = \begin{bmatrix} p_{x,l} \\ p_{y,l} \\ p_{z,l} \end{bmatrix} \tag{2.45}$$

with l the particle index. Therefore, defining each semiclassical Hamiltonian eigenstate via such coordinates and momenta and treating the phase space tiny volumes as true differential volumes, we can write the canonical partition function of the semiclassical ideal gas system as

$$\begin{aligned} Q &\cong \frac{1}{N!}\Pi_{l=1}^{N} \int e^{-\beta\left(\frac{p_{x,l}^2}{2m}+\frac{p_{y,l}^2}{2m}+\frac{p_{z,l}^2}{2m}\right)} \frac{d\boldsymbol{p}_l d\boldsymbol{r}_l}{\delta^3} \\ &= \frac{1}{N!}\Pi_{l=1}^{N}\left(\frac{L_x}{\delta}\int_{-\infty}^{\infty} e^{-\beta\frac{p_{x,l}^2}{2m}}\, dp_{x,l}\right)\left(\frac{L_y}{\delta}\int_{-\infty}^{\infty} e^{-\beta\frac{p_{y,l}^2}{2m}}\, dp_{y,l}\right)\left(\frac{L_z}{\delta}\int_{-\infty}^{\infty} e^{-\beta\frac{p_{z,l}^2}{2m}}\, dp_{z,l}\right) \\ &= \frac{V^N}{\delta^{3N} N!}\left[2\pi m/\beta\right]^{3N/2} \end{aligned} \tag{2.46}$$

where $1/\delta^3$ is the wave packet density within the single particle phase space (i.e., the density of semiclassical eigenstates in the full phase space is δ^{-3N}) and $N!$ is again the correction for identical particle permutations. Comparison between Eqs. 2.46 and 2.43 plainly shows that $\delta = h$ and hence, for whatever system considered, the semiclassical eigenstate density in the phase space defined by D semiclassical degrees of freedom must be h^{-D}.

2.3.3 *Fundamental Statistical Thermodynamics*

When using $\beta = 1/(k_B T)$ and defining with $F = -1/\beta \ln Q = -k_B T \ln Q$ the canonical free energy, we readily obtain that in Eqs. 2.23–2.25 the canonical free energy $F(N, V, T)$ corresponds exactly to the Helmholtz free energy in Eqs. 2.26–2.28. Moreover, from

$$\begin{aligned} \lim_{T\to 0} F(N, V, T) &= \lim_{T\to 0} -k_B T \ln Q = \mathcal{U}_0 - \lim_{T\to 0} k_B T \ln\Big(\sum_j e^{-\frac{\mathcal{U}_j - \mathcal{U}_0}{k_B T}}\Big) \\ &= \mathcal{U}_0 = \lim_{T\to 0} A(N, V, T) \end{aligned} \tag{2.47}$$

with $\mathcal{U}_0(N, V)$ the system ground eigenstate energy, it follows that $F = -k_B T \ln Q(N, V, T)$ share with the Helmholtz free energy the derivatives in the thermodynamic state variables N, V, T as well as the zero temperature limit, thus being necessarily identical to the Helmholtz free energy for whatever N, V, T condition. We can therefore write

$$A(N, V, T) = -k_B T \ln Q \tag{2.48}$$

$$S(N, V, T) = \frac{U - A}{T} = \frac{U}{T} + k_B \ln Q \tag{2.49}$$

$$G(N, V, T) = A + pV = -k_B T \ln Q + pV \tag{2.50}$$

$$H(N, V, T) = U + pV \tag{2.51}$$

$$p(N, V, T) = k_B T \left(\frac{\partial \ln Q}{\partial V}\right)_{N,T} \tag{2.52}$$

$$U(N, V, T) = -k_B \left(\frac{\partial \ln Q}{\partial 1/T}\right)_{N,V} \tag{2.53}$$

$$\mu_l(N, V, T) = -k_B T \left(\frac{\partial \ln Q}{\partial N_l}\right)_{N_l,V,T} \tag{2.54}$$

with S the entropy, G the Gibbs free energy, H the enthalpy of the system and $\mu_l = (\partial A/\partial N_l)_{N_l,V,T} = (\partial G/\partial N_l)_{N_l,p,T}$ the lth species chemical potential, all as obtained by the canonical ensemble statistics. Equations 2.48–2.54 can then provide (using also their derivatives) any thermodynamic property in terms of the canonical partition function, e.g., the isochoric heat capacity $C_V(N, V, T) = (\partial U/\partial T)_{N,V}$ is obtained by

$$\begin{aligned} C_V(N, V, T) &= \left(\frac{\partial U}{\partial T}\right)_{N,V} = -\frac{1}{k_B T^2}\left(\frac{\partial U}{\partial \beta}\right)_{N,V} \\ &= -\frac{1}{k_B T^2}\left\{-\frac{\sum_j e^{-\beta \mathcal{U}_j}\mathcal{U}_j^2}{Q} + \frac{(\sum_j e^{-\beta \mathcal{U}_j}\mathcal{U}_j)^2}{Q^2}\right\} \\ &= \frac{\langle \mathcal{U}_j^2\rangle - \langle \mathcal{U}_j\rangle^2}{k_B T^2} \end{aligned} \tag{2.55}$$

with the angle brackets indicating the canonical ensemble average and obviously $\langle \mathcal{U}_j\rangle = U(N, V, T)$.

2.3.4 A Canonical Ensemble Special Case: The Microcanonical Ensemble

From the definition and properties of the canonical ensemble and related partition function we can derive the equilibrium distribution for a quasi-closed system with quasi-stationary Hamiltonian, corresponding to a special case of the canonical ensemble. In fact, we can always express the canonical partition function via

$$Q(N,V,T) = \sum_j e^{-\beta \mathcal{U}_j} = \sum_{\mathcal{U}} e^{-\beta \mathcal{U}} \Omega(N,V,\mathcal{U}) \tag{2.56}$$

where $\mathcal{U}$ are the Hamiltonian eigenvalues and $\Omega(N,V,\mathcal{U})$ provides the number of Hamiltonian eigenstates with energy identical or negligibly different from $\mathcal{U}$ (i.e., $|\mathcal{U}_j - \mathcal{U}| \leq \delta_{\mathcal{U}}$ with $\delta_{\mathcal{U}} > 0$ a tiny energy interval). Let us consider a statistical ensemble made of canonical-like systems with however a virtually fixed energy $\mathcal{U} \cong \mathcal{U}_{ref}$ (i.e., the microcanonical ensemble), thus corresponding to the canonical ensemble when constraining the system energy at $\mathcal{U} \cong \mathcal{U}_{ref}$. From Eq. 2.56 the canonical partition function and eigenstate probability distribution once allowing only the eigenstates with $\mathcal{U} \cong \mathcal{U}_{ref}$ is

$$Q(N,V,\mathcal{U}_{ref}) = e^{-\beta \mathcal{U}_{ref}} \Omega(N,V,\mathcal{U}_{ref}) \tag{2.57}$$

$$\rho_j = \frac{e^{-\beta \mathcal{U}_{ref}} \Gamma_j}{Q(N,V,\mathcal{U}_{ref})} = \frac{\Gamma_j}{\Omega(N,V,\mathcal{U}_{ref})} \tag{2.58}$$

where Γ_j is the step function

$$|\mathcal{U}_j - \mathcal{U}_{ref}| \leq \delta_{\mathcal{U}}$$
$$\Gamma_j = 1$$
$$|\mathcal{U}_j - \mathcal{U}_{ref}| > \delta_{\mathcal{U}}$$
$$\Gamma_j = 0$$

and obviously $\mathcal{U}_{ref}$ is the corresponding equilibrium average energy, with then Eq. 2.58 providing the microcanonical ensemble eigenstate distribution. Note that the obtained microcanonical statistics, i.e., a constant probability for all the Hamiltonian eigenstates with eigenvalue virtually identical to $\mathcal{U}_{ref}$, is in full agreement with the classical mechanical microcanonical ensemble.

2.3.5 *The Microcanonical Entropy*

Realizing that the quantum energy $\mathcal{U}$ can be treated as a continuous variable as the volume size and huge number of particles of a macroscopic system make its Hamiltonian eigenvalues a quasi-continuous set of energies, we can treat $e^{-\beta \mathcal{U}} \Omega$ as a continuous function of $\mathcal{U}$, with $e^{-\beta \mathcal{U}}$ steeply decreasing and Ω steeply increasing as a function of $\mathcal{U}$. Therefore, $e^{-\beta \mathcal{U}} \Omega$ is a highly peaked function with the energy corresponding to the function maximum (i.e., $\mathcal{U}_M$) provided by

$$\left(\frac{\partial e^{-\beta\mathcal{U}}\Omega}{\partial\mathcal{U}}\right)_{N,V} = -\beta e^{-\beta\mathcal{U}}\Omega + e^{-\beta\mathcal{U}}\left(\frac{\partial\Omega}{\partial\mathcal{U}}\right)_{N,V} = 0 \quad (2.59)$$

and thus

$$k_B\left(\frac{\partial\Omega/\partial\mathcal{U}}{\Omega}\right)_{N,V} = \frac{1}{T} \quad (2.60)$$

at $\mathcal{U} = \mathcal{U}_M$. Furthermore, considering that due to the highly peaked $e^{-\beta\mathcal{U}}\Omega$ behavior $\mathcal{U}_M \cong U$, with U the system equilibrium average energy in the canonical ensemble, when comparing Eq. 2.60 with the general thermodynamic relation $1/T = (\partial S/\partial U)_{N,V}$ we can express the system entropy $S(N, V, T)$ in the canonical ensemble via

$$S \cong S_U = k_B \ln \Omega(N, V, U) \quad (2.61)$$

and hence more in general

$$S_{\mathcal{U}} = k_B \ln \Omega(N, V, \mathcal{U}) \quad (2.62)$$

providing the system microcanonical entropy at energy $\mathcal{U}$ (i.e., the system entropy in the microcanonical ensemble where $U = \mathcal{U}$).

Using Eq. 2.62 into Eq. 2.56 we obtain (indicating with U the system equilibrium average energy in the canonical ensemble)

$$\begin{aligned} Q(N, V, T) &= \sum_{\mathcal{U}} e^{-\beta(\mathcal{U}-TS_{\mathcal{U}})} \\ &= e^{-\beta(U-TS_U)}\left\{1 + \sum_{\mathcal{U}\neq U} e^{-\beta(\Delta\mathcal{U}-T\Delta S_{\mathcal{U}})}\right\} \quad (2.63) \\ \Delta\mathcal{U} &= \mathcal{U} - U \quad (2.64) \\ \Delta S_{\mathcal{U}} &= S_{\mathcal{U}} - S_U \quad (2.65) \end{aligned}$$

providing for the canonical ensemble Helmholtz free energy

$$\begin{aligned} A(N, V, T) &= -k_B T \ln Q(N, V, T) \\ &= U - TS_U - k_B T \ln\left[1 + \sum_{\mathcal{U}\neq U} e^{-\beta(\Delta\mathcal{U}-T\Delta S_{\mathcal{U}})}\right] \\ &= U - T\left\{S_U + k_B \ln\left[1 + \sum_{\mathcal{U}\neq U} e^{-\beta(\Delta\mathcal{U}-T\Delta S_{\mathcal{U}})}\right]\right\} \quad (2.66) \end{aligned}$$

thus readily showing that the system entropy in the canonical ensemble is given by

$$S(N, V, T) = S_U + k_B \ln\left[1 + \sum_{\mathcal{U}\neq U} e^{-\beta(\Delta\mathcal{U}-T\Delta S_{\mathcal{U}})}\right] \quad (2.67)$$

Finally, comparing the last equation with Eq. 2.61 we necessarily have

$$k_B \ln\left\{1 + \sum_{\mathcal{U}\neq U} e^{-\beta(\Delta\mathcal{U} - T\Delta S_{\mathcal{U}})}\right\} \cong 0 \tag{2.68}$$

and hence

$$A(N, V, T) \cong U(N, V, T) - k_B T \ln \Omega(N, V, U) \tag{2.69}$$

the approximation becoming exact as the system number of molecules tends to infinity. In practice, the inaccuracy of such an approximation can be fully disregarded for any system equal or larger than one elementary system (i.e., a thermodynamic system).

2.4 Non-canonical Ensembles

As it is described in the previous sections, the canonical ensemble is defined by systems at fixed volume and chemical composition in thermal equilibrium with their environment (i.e., the thermal bath), thus characterized by the energy fluctuations and constrained volume and chemical species amounts. In order to reconstruct the statistics of systems with the volume or the chemical composition fluctuating, it is worth to introduce the typical non-canonical ensembles.

2.4.1 The Isothermal–Isobaric Ensemble

Let us consider a huge canonical system embedding a macroscopic system in thermal equilibrium with its environment and at fixed chemical composition with, however, a fluctuating volume (i.e., the macroscopic system ideal box does not allow any matter exchange but does allow the volume exchange like for the heat) thus resulting also in a basic equilibrium with its canonical environment (i.e., the much larger complementary part of the huge canonical system). We can obviously consider such macroscopic system and its canonical environment as statistically independent and hence at each macroscopic system volume V' we can express the Helmholtz free energy of the whole canonical system via $A(N, V', T) + A_{env}(N_{env}, V_{env}, T)$ where

$$A(N, V', T) = -k_B T \ln Q(N, V', T) \tag{2.70}$$

$$A_{env}(N_{env}, V_{env}, T) = -k_B T \ln Q_{env}(N_{env}, V_{env}, T) \tag{2.71}$$

are the macroscopic system and its canonical environment free energies, as expressed by the corresponding canonical partition functions, N, N_{env} are the fixed chemical compositions of the macroscopic system and its canonical environment and $V_{env} =$

$V_w - V'$ is the canonical environment volume with V_w the whole canonical system fixed volume. Therefore, the whole canonical system partition function Q_w can be written as

$$Q_w(N, N_{env}, V_w, T) = \sum_{V'} e^{-\beta[A(N,V',T)+A_{env}(N_{env},V_w-V',T)]} \tag{2.72}$$

with V' running over all the accessible macroscopic system volumes (a quasi-continuous set of volumes). Considering that the huge canonical environment can hardly change its free energy as a function of the macroscopic system volume variations, expanding to first order in V_{env} around $V_{env} = V_w$ (i.e., $V' = 0$), we have

$$\begin{aligned} A_{env}(N_{env}, V_w - V', T) &= A_{env}(N_{env}, V_w, T) - \left(\frac{\partial A_{env}}{\partial V_{env}}\right)_{V_{env}=V_w} V' \\ &= A_{env}(N_{env}, V_w, T) + pV' \end{aligned} \tag{2.73}$$

$$Q_{env}(N_{env}, V_w - V', T) = e^{-\beta A_{env}(N_{env},V_w,T)} \, e^{-\beta p V'} \tag{2.74}$$

with $(\partial A_{env}/\partial V_{env})_{V_{env}=V_w}$ the partial derivative $(\partial A_{env}/\partial V_{env})_{N_{env},T}$ as obtained at $V_{env} = V_w$, and hence

$$Q_w(N, N_{env}, V_w, T) = e^{-\beta A_{env}(N_{env},V_w,T)} \sum_{V'} e^{-\beta[A(N,V',T)+pV']} \tag{2.75}$$

where $p = -(\partial A_{env}/\partial V_{env})_{V_{env}=V_w}$ is the equilibrium pressure of the canonical environment (due to its huge size, the canonical environment equilibrium pressure is independent of the macroscopic system volume). From the last equation we readily obtain the probability of a given jth Hamiltonian eigenstate (with eigenvalue $\mathcal{U}_j$) of the macroscopic system at volume V' by

$$\begin{aligned} \rho_j(V') &= \frac{e^{-\beta \mathcal{U}_j(N,V')}}{Q_w} Q_{env}(N_{env}, V_w - V', T) \\ &= \frac{e^{-\beta[\mathcal{U}_j(N,V')+pV']}}{\sum_{V'} e^{-\beta[A(N,V',T)+pV']}} \end{aligned} \tag{2.76}$$

$$e^{-\beta A(N,V',T)} = Q(N, V', T) = \sum_j e^{-\beta \mathcal{U}_j(N,V')} \tag{2.77}$$

By analogy with the canonical ensemble distribution, from Eq. 2.76 we can define the isothermal–isobaric partition function via

$$\begin{aligned} \Delta(N, p, T) &= \sum_{V'} e^{-\beta[A(N,V',T)+pV']} \\ &= \sum_{V'} \sum_j e^{-\beta[\mathcal{U}_j(N,V')+pV']} \end{aligned} \tag{2.78}$$

and hence the isothermal–isobaric free energy by

$$F(N, p, T) = -k_B T \ln \Delta(N, p, T) \tag{2.79}$$

From Eqs. 2.76–2.79 we then obtain

$$\begin{aligned}
\left(\frac{\partial F}{\partial p}\right)_{N,T} &= \frac{\sum_{V'} \sum_j e^{-\beta[\mathcal{U}_j(N,V')+pV']}\, V'}{\sum_{V'} e^{-\beta[A(N,V',T)+pV']}} = \frac{\sum_{V'} e^{-\beta[A(N,V',T)+pV']}\, V'}{\sum_{V'} e^{-\beta[A(N,V',T)+pV']}} \\
&= \langle V' \rangle = V = \left(\frac{\partial G}{\partial p}\right)_{N,T}
\end{aligned} \tag{2.80}$$

$$\begin{aligned}
\left(\frac{\partial F/T}{\partial 1/T}\right)_{N,p} &= \frac{\sum_{V'} \sum_j e^{-\beta[\mathcal{U}_j(N,V')+pV']}\, [\mathcal{U}_j(N, V') + pV']}{\sum_{V'} e^{-\beta[A(N,V',T)+pV']}} \\
&= \frac{\sum_{V'} e^{-\beta[A(N,V',T)+pV']}\, [U(N, V', T) + pV']}{\sum_{V'} e^{-\beta[A(N,V',T)+pV']}} \\
&= \langle U(N, V', T) \rangle + pV = U(N, p, T) + pV \\
&= H(N, p, T) = \left(\frac{\partial G/T}{\partial 1/T}\right)_{N,p}
\end{aligned} \tag{2.81}$$

$$\begin{aligned}
\left(\frac{\partial F}{\partial N_l}\right)_{N_l,p,T} &= \frac{\sum_{V'} \sum_j e^{-\beta[\mathcal{U}_j(N,V')+pV']}\, (\partial \mathcal{U}_j/\partial N_l)_{N_l,V'}}{\sum_{V'} e^{-\beta[A(N,V',T)+pV']}} \\
&= \frac{\sum_{V'} e^{-\beta[A(N,V',T)+pV']}\, \mu_l(N, V', T)}{\sum_{V'} e^{-\beta[A(N,V',T)+pV']}} \\
&= \langle \mu_l(N, V', T) \rangle = \mu_l(N, p, T) = \left(\frac{\partial G}{\partial N_l}\right)_{N_l,p,T}
\end{aligned} \tag{2.82}$$

$$U(N, V', T) = \frac{\sum_j e^{-\beta \mathcal{U}_j(N,V')}\, \mathcal{U}_j}{Q(N, V', T)} \tag{2.83}$$

$$\mu_l(N, V', T) = \frac{\sum_j e^{-\beta \mathcal{U}_j(N,V')}\, (\partial \mathcal{U}_j/\partial N_l)_{N,V'}}{Q(N, V', T)} \tag{2.84}$$

with $V(N, p, T)$, $H(N, p, T)$, $\mu_l(N, p, T)$ the average volume, the enthalpy and the lth species chemical potential within the isothermal–isobaric ensemble (the angle brackets indicating the isothermal–isobaric ensemble average), and

$$\begin{aligned}
\lim_{T\to 0} F(N, p, T) &= H_{min} - \lim_{T\to 0} k_B T \ln\left[\sum_{V'} \sum_j e^{-\beta(\mathcal{U}_j+pV'-H_{min})}\right] \\
&= H_{min} = \lim_{T\to 0} G(N, p, T)
\end{aligned} \tag{2.85}$$

where H_{min} is the minimum value of $\mathcal{U}_0(N, V') + pV'$ as obtained by $-(\partial \mathcal{U}_0/\partial V')_N = p$ (i.e., provided by the V' value where the ground eigenstate pressure is identical to the equilibrium pressure).

From Eqs. 2.80–2.85 we then obtain $F(\boldsymbol{N}, p, T) = G(\boldsymbol{N}, p, T)$ showing that in the isothermal–isobaric ensemble we have

$$G(\boldsymbol{N}, p, T) = -k_B T \ln \Delta \tag{2.86}$$

$$S(\boldsymbol{N}, p, T) = \frac{H(\boldsymbol{N}, p, T) - G(\boldsymbol{N}, p, T)}{T} = \frac{H(\boldsymbol{N}, p, T)}{T} + k_B \ln \Delta \tag{2.87}$$

$$V(\boldsymbol{N}, p, T) = \langle V' \rangle = -k_B T \left(\frac{\partial \ln \Delta}{\partial p} \right)_{\boldsymbol{N},T} \tag{2.88}$$

$$H(\boldsymbol{N}, p, T) = U(\boldsymbol{N}, p, T) + pV = -k_B \left(\frac{\partial \ln \Delta}{\partial 1/T} \right)_{\boldsymbol{N},p} \tag{2.89}$$

$$\mu_l(\boldsymbol{N}, p, T) = \langle \mu_l(\boldsymbol{N}, V', T) \rangle = -k_B T \left(\frac{\partial \ln \Delta}{\partial N_l} \right)_{N_l,p,T} \tag{2.90}$$

$$A(\boldsymbol{N}, p, T) = G(\boldsymbol{N}, p, T) - pV \tag{2.91}$$

$$U(\boldsymbol{N}, p, T) = \langle U(\boldsymbol{N}, V', T) \rangle \tag{2.92}$$

Equations 2.86–2.92 can then provide (using also their derivatives) any thermodynamic property in terms of the isothermal–isobaric partition function, e.g., the isobaric heat capacity $C_p(\boldsymbol{N}, p, T) = (\partial H/\partial T)_{\boldsymbol{N},p}$ is obtained by

$$\begin{aligned}
C_p(\boldsymbol{N}, p, T) &= \left(\frac{\partial H}{\partial T} \right)_{\boldsymbol{N},p} = -\frac{1}{k_B T^2} \left(\frac{\partial H}{\partial \beta} \right)_{\boldsymbol{N},p} \\
&= -\frac{1}{k_B T^2} \left\{ -\frac{\sum_{V'} \sum_j e^{-\beta(\mathcal{U}_j + pV')} (\mathcal{U}_j + pV')^2}{\Delta} + \frac{\left[\sum_{V'} \sum_j e^{-\beta(\mathcal{U}_j + pV')} (\mathcal{U}_j + pV') \right]^2}{\Delta^2} \right\} \\
&= \frac{\langle (\mathcal{U}_j + pV')^2 \rangle - \langle \mathcal{U}_j + pV' \rangle^2}{k_B T^2} = \frac{\langle [\mathcal{U}_j - \langle \mathcal{U}_j \rangle + p(V' - V)]^2 \rangle}{k_B T^2} \\
&= \frac{\langle (\mathcal{U}_j - \langle \mathcal{U}_j \rangle)^2 \rangle}{k_B T^2} + p^2 \frac{\langle (V' - V)^2 \rangle}{k_B T^2} + 2p \frac{\langle (\mathcal{U}_j - \langle \mathcal{U}_j \rangle)(V' - V) \rangle}{k_B T^2}
\end{aligned} \tag{2.93}$$

with the angle brackets indicating the isothermal–isobaric ensemble average and obviously

$$\langle \mathcal{U}_j + pV' \rangle = \langle \mathcal{U}_j \rangle + pV = H(\boldsymbol{N}, p, T) \tag{2.94}$$

From Eqs. 2.78, 2.86 and 2.87 we can express the Gibbs free energy and the entropy as

$$G(\boldsymbol{N}, p, T) = A(\boldsymbol{N}, V, T) + pV - k_B T \ln \left\{ 1 + \sum_{V' \neq V} e^{-\beta[\Delta A(\boldsymbol{N}, V', T) + p\Delta V']} \right\} \tag{2.95}$$

$$S(\boldsymbol{N}, p, T) = \frac{H(\boldsymbol{N}, p, T) - A(\boldsymbol{N}, V, T) - pV}{T}$$

$$
\begin{aligned}
&+ k_B \ln \left\{ 1 + \sum_{V' \neq V} e^{-\beta[\Delta A(N,V',T) + p\Delta V']} \right\} \\
&= \frac{\langle U(N, V', T) \rangle - U(N, V, T)}{T} + S(N, V, T) \\
&+ k_B \ln \left\{ 1 + \sum_{V' \neq V} e^{-\beta[\Delta A(N,V',T) + p\Delta V']} \right\} \qquad (2.96)
\end{aligned}
$$

$$
\Delta A(N, V', T) = A(N, V', T) - A(N, V, T) \qquad (2.97)
$$

$$
\Delta V' = V' - V \qquad (2.98)
$$

where again the angle brackets indicate averaging in the isothermal–isobaric ensemble and for whatever thermodynamic property X, i.e., the equilibrium property X, with the notation $X(N, p, T)$ we mean the property as obtained within the isothermal–isobaric ensemble while when using the notation $X(N, V', T)$ or $X(N, V, T)$ we indicate the same property as provided by the canonical ensemble.

Finally, considering as usual that $e^{-\beta[A(N,V',T)+pV']}$ is highly peaked and thus its maximum is located at the volume $V_M \cong \langle V' \rangle = V$ we have

$$
U(N, p, T) = \langle U(N, V', T) \rangle \cong U(N, V, T) \qquad (2.99)
$$

$$
k_B T \ln \left[1 + \sum_{V' \neq V} e^{-\beta[\Delta A(N,V',T) + pV']} \right] \cong 0 \qquad (2.100)
$$

and realizing that at $V' = V_M \cong V$

$$
p(N, V', T) = -(\partial A / \partial V')_{N,T} = p \qquad (2.101)
$$

that is $p(N, V, T) \cong p$, we can write

$$
G(N, p, T) \cong A(N, V, T) + pV \cong G(N, V, T) \qquad (2.102)
$$

$$
S(N, p, T) \cong S(N, V, T) \qquad (2.103)
$$

$$
H(N, p, T) \cong U(N, V, T) + pV \cong H(N, V, T) \qquad (2.104)
$$

$$
\mu_l(N, p, T) = \langle \mu_l(N, V', T) \rangle \cong \mu_l(N, V, T) \qquad (2.105)
$$

the approximations becoming fully accurate for any system equal or larger than one elementary system (i.e., a thermodynamic system), thus showing that the isothermal–isobaric and canonical ensembles provide fully equivalent thermodynamics.

2.4.2 *The Grand-Canonical Ensemble*

Similarly to the previous subsection, we can consider a huge canonical system embedding a macroscopic system in thermal equilibrium with its environment and at fixed volume V with, however, a fluctuating chemical composition (i.e., the macroscopic system rigid box does allow matter exchange) thus resulting also in matter equilibrium with its canonical environment (i.e., the much larger complementary part of the huge canonical system). We can obviously consider again such macroscopic system and its canonical environment as statistically independent and hence at each macroscopic system chemical composition $\boldsymbol{N}'$ we can express the Helmholtz free energy of the whole canonical system via $A(\boldsymbol{N}', V, T) + A_{env}(\boldsymbol{N}_{env}, V_{env}, T)$ where

$$A(\boldsymbol{N}', V, T) = -k_B T \ln Q(\boldsymbol{N}', V, T) \tag{2.106}$$

$$A_{env}(\boldsymbol{N}_{env}, V_{env}, T) = -k_B T \ln Q_{env}(\boldsymbol{N}_{env}, V_{env}, T) \tag{2.107}$$

are the macroscopic system and its canonical environment free energies, as expressed by the corresponding canonical partition functions, $\boldsymbol{N}_{env} = \boldsymbol{N}_w - \boldsymbol{N}'$ is the chemical composition of the canonical environment (with $\boldsymbol{N}_w$ the fixed chemical composition of the whole, huge, canonical system) and V_{env} is the fixed canonical environment volume. Therefore, the whole canonical system partition function Q_w can be written as

$$Q_w(\boldsymbol{N}', \boldsymbol{N}_{env}, V_w, T) = \sum_{\boldsymbol{N}'} e^{-\beta[A(\boldsymbol{N}',V,T)+A_{env}(\boldsymbol{N}_w-\boldsymbol{N}',V_{env},T)]}$$

$$\cong \sum_{N_1'=0}^{\infty}\sum_{N_2'=0}^{\infty}\cdots\sum_{N_l'=0}^{\infty} e^{-\beta[A(\boldsymbol{N}',V,T)+A_{env}(\boldsymbol{N}_w-\boldsymbol{N}',V_{env},T)]} \tag{2.108}$$

with each N_l' running over all the accessible macroscopic system lth chemical species amounts, including a virtually infinite number of molecules due to the huge molecular amount possibly present into a macroscopic size system. Considering that the enormous canonical environment can hardly change its free energy as a function of the macroscopic system chemical composition variations, expanding to first order in $\boldsymbol{N}_{env}$ around $\boldsymbol{N}_{env} = \boldsymbol{N}_w$ (i.e., $\boldsymbol{N}' = \boldsymbol{0}$), we have

$$A_{env}(\boldsymbol{N}_w - \boldsymbol{N}', V_{env}, T) = A_{env}(\boldsymbol{N}_w, V_{env}, T) - \sum_l \left(\frac{\partial A_{env}}{\partial N_{env,l}}\right)_{\boldsymbol{N}_{env}=\boldsymbol{N}_w} N_l'$$

$$= A_{env}(\boldsymbol{N}_w, V_{env}, T) - \sum_l \mu_l N_l' \tag{2.109}$$

$$Q_{env}(\boldsymbol{N}_w - \boldsymbol{N}', V_{env}, T) = e^{-\beta A_{env}(\boldsymbol{N}_w,V_{env},T)}\, e^{\beta \sum_l \mu_l N_l'} \tag{2.110}$$

with $(\partial A_{env}/\partial N_{env,l})_{\boldsymbol{N}_{env}=\boldsymbol{N}_w}$ the partial derivative $(\partial A_{env}/\partial N_{env,l})_{\boldsymbol{N}_{env,l},V_{env},T}$ as obtained at $\boldsymbol{N}_{env} = \boldsymbol{N}_w$ ($\boldsymbol{N}_{env,l}$ being clearly the vector of the environment molecular amounts once removing $N_{env,l}$), and hence

$$Q_w(\boldsymbol{N}, \boldsymbol{N}_{env}, V_w, T) = e^{-\beta A_{env}(\boldsymbol{N}_w,V_{env},T)} \sum_{\boldsymbol{N}'} e^{-\beta[A(\boldsymbol{N}',V,T)-\sum_l \mu_l N_l']} \tag{2.111}$$

where $\mu_l = (\partial A_{env}/\partial N_{env,l})_{N_{env}=N_w}$ is the lth species equilibrium chemical potential in the canonical environment (due to its huge size, such canonical environment chemical potentials are independent of the macroscopic system composition). From the last equation we readily obtain the probability of a given jth Hamiltonian eigenstate (with eigenvalue $\mathcal{U}_j$) of the macroscopic system at chemical composition $\mathbf{N}'$ by

$$\begin{aligned} \rho_j(\mathbf{N}') &= \frac{e^{-\beta \mathcal{U}_j(\mathbf{N}',V)}}{Q_w} Q_{env}(\mathbf{N}_w - \mathbf{N}', V_{env}, T) \\ &= \frac{e^{-\beta[\mathcal{U}_j(\mathbf{N}',V) - \sum_l \mu_l N_l']}}{\sum_{\mathbf{N}'} e^{-\beta[A(\mathbf{N}',V,T) - \sum_l \mu N_l']}} \end{aligned} \tag{2.112}$$

$$e^{-\beta A(\mathbf{N}',V,T)} = Q(\mathbf{N}', V, T) = \sum_j e^{-\beta \mathcal{U}_j(\mathbf{N}',V)} \tag{2.113}$$

By analogy with the canonical and isothermal–isobaric ensemble distributions, we can then define the grand-canonical partition function via

$$\begin{aligned} Z(\boldsymbol{\mu}, V, T) &= \sum_{\mathbf{N}'} e^{-\beta[A(\mathbf{N}',V,T) - \sum_l \mu_l N_l']} \\ &= \sum_{\mathbf{N}'} \sum_j e^{-\beta[\mathcal{U}_j(\mathbf{N}',V) - \sum_l \mu_l N_l']} \end{aligned} \tag{2.114}$$

and hence the grand-canonical free energy by

$$F(\boldsymbol{\mu}, V, T) = -k_B T \ln Z(\boldsymbol{\mu}, V, T) \tag{2.115}$$

with $\boldsymbol{\mu}$ the vector of the equilibrium chemical potentials (i.e., $\mu_1, \mu_2, \ldots, \mu_l, \ldots$).

From Eqs. 2.112–2.115 we then obtain (introducing the thermodynamic potential $\Phi_G = A - G = -pV$)

$$\begin{aligned} \left(\frac{\partial F}{\partial V}\right)_{\boldsymbol{\mu},T} &= \frac{\sum_{\mathbf{N}'} \sum_j e^{-\beta[\mathcal{U}_j(\mathbf{N}',V) - \sum_l \mu_l N_l']} (\partial \mathcal{U}_j/\partial V)_{\mathbf{N}',T}}{\sum_{\mathbf{N}'} e^{-\beta[A(\mathbf{N}',V,T) - \sum_l \mu_l N_l']}} \\ &= -\frac{\sum_{\mathbf{N}'} e^{-\beta[A(\mathbf{N}',V,T) - \sum_l \mu_l N_l']} p(\mathbf{N}', V, T)}{\sum_{\mathbf{N}'} e^{-\beta[A(\mathbf{N},V',T) - \sum_l \mu_l N_l']}} \\ &= -\langle p(\mathbf{N}', V, T)\rangle = -p(\boldsymbol{\mu}, V, T) = \left(\frac{\partial \Phi_G}{\partial V}\right)_{\boldsymbol{\mu},T} \end{aligned} \tag{2.116}$$

$$\begin{aligned} \left(\frac{\partial F/T}{\partial 1/T}\right)_{\boldsymbol{\mu},V} &= \frac{\sum_{\mathbf{N}'} \sum_j e^{-\beta[\mathcal{U}_j(\mathbf{N}',V) - \sum_l \mu_l N_l']} [\mathcal{U}_j(\mathbf{N}', V) - \sum_l \mu_l N_l']}{\sum_{\mathbf{N}'} e^{-\beta[A(\mathbf{N}',V,T) - \sum_l \mu_l N_l']}} \\ &= \frac{\sum_{\mathbf{N}'} e^{-\beta[A(\mathbf{N}',V,T) - \sum_l \mu_l N_l']} [U(\mathbf{N}', V, T) - \sum_l \mu_l N_l']}{\sum_{\mathbf{N}'} e^{-\beta[A(\mathbf{N}',V,T) - \sum_l \mu_l N_l']}} \end{aligned}$$

$$
\begin{aligned}
&= \langle U(\mathbf{N}', V, T)\rangle - \sum_l \mu_l \langle N_l'\rangle = U(\boldsymbol{\mu}, V, T) - \sum_l \mu_l N_l \\
&= U(\boldsymbol{\mu}, V, T) - G(\boldsymbol{\mu}, V, T) = D(\boldsymbol{\mu}, V, T) \\
&= \left(\frac{\partial \Phi_G / T}{\partial 1/T}\right)_{\boldsymbol{\mu}, V}
\end{aligned} \tag{2.117}
$$

$$
\begin{aligned}
\left(\frac{\partial F}{\partial \mu_l}\right)_{\boldsymbol{\mu}_l, V, T} &= -\frac{\sum_{\mathbf{N}'} \sum_j e^{-\beta[\mathcal{U}_j(\mathbf{N}', V) - \sum_l \mu_l N_l']} N_l'}{\sum_{\mathbf{N}'} e^{-\beta[A(\mathbf{N}', V, T) - \sum_l \mu_l N_l']}} \\
&= -\frac{\sum_{\mathbf{N}'} e^{-\beta[A(\mathbf{N}', V, T) - \sum_l \mu_l N_l']} N_l'}{\sum_{\mathbf{N}'} e^{-\beta[A(\mathbf{N}', V, T) - \sum_l \mu_l N_l']}} \\
&= -\langle N_l'\rangle = -N_l = \left(\frac{\partial \Phi_G}{\partial \mu_l}\right)_{\boldsymbol{\mu}_l, V, T}
\end{aligned} \tag{2.118}
$$

$$
p(\mathbf{N}', V, T) = -\frac{\sum_j e^{-\beta \mathcal{U}_j(\mathbf{N}', V)} (\partial \mathcal{U}_j / \partial V)_{\mathbf{N}', T}}{Q(\mathbf{N}', V, T)} \tag{2.119}
$$

$$
U(\mathbf{N}', V, T) = \frac{\sum_j e^{-\beta \mathcal{U}_j(\mathbf{N}', V)} \mathcal{U}_j}{Q(\mathbf{N}', V, T)} \tag{2.120}
$$

with $p(\boldsymbol{\mu}, V, T), D(\boldsymbol{\mu}, V, T), N_l(\boldsymbol{\mu}, V, T)$ the average pressure, the grand-canonical enthalpy and the lth chemical species average number of molecules within the grand-canonical ensemble (the angle brackets indicating the grand-canonical ensemble average and $\boldsymbol{\mu}_l$ the vector of the equilibrium chemical potentials when removing μ_l). Furthermore, we have

$$
\begin{aligned}
\lim_{T\to 0} F(\boldsymbol{\mu}, V, T) &= D_{min} - \lim_{T\to 0} k_B T \ln\Big[\sum_{\mathbf{N}'}\sum_j e^{-\beta(\mathcal{U}_j - \sum_l \mu_l N_l' - D_{min})}\Big] \\
&= D_{min} = \lim_{T\to 0} \Phi_G(\boldsymbol{\mu}, V, T)
\end{aligned} \tag{2.121}
$$

where D_{min} is the minimum value of $\mathcal{U}_0(\mathbf{N}', V) - \sum_l \mu_l N_l'$ as obtained by the $\mathbf{N}'$ molecular amounts providing $(\partial \mathcal{U}_0 / \partial N_l')_{\mathbf{N}_l', V} = \mu_l$ ($\mathbf{N}_l'$ is clearly the vector of the system molecular amounts once removing N_l').

From Eqs. 2.116–2.121 we then obtain $F(\boldsymbol{\mu}, V, T) = \Phi_G(\boldsymbol{\mu}, V, T) = -p(\boldsymbol{\mu}, V, T)V$ showing that for the grand-canonical ensemble we can write

$$
\begin{aligned}
\Phi_G(\boldsymbol{\mu}, V, T) &= A(\boldsymbol{\mu}, V, T) - G(\boldsymbol{\mu}, V, T) \\
&= D(\boldsymbol{\mu}, V, T) - TS(\boldsymbol{\mu}, V, T) = -k_B T \ln Z
\end{aligned} \tag{2.122}
$$

$$
\begin{aligned}
S(\boldsymbol{\mu}, V, T) &= \frac{D(\boldsymbol{\mu}, V, T) - \Phi_G(\boldsymbol{\mu}, V, T)}{T} \\
&= \frac{D(\boldsymbol{\mu}, V, T)}{T} + k_B \ln Z
\end{aligned} \tag{2.123}
$$

$$
p(\boldsymbol{\mu}, V, T) = \langle p(\mathbf{N}', V, T)\rangle = k_B T \left(\frac{\partial \ln Z}{\partial V}\right)_{\boldsymbol{\mu}, T} \tag{2.124}
$$

$$D(\boldsymbol{\mu}, V, T) = U(\boldsymbol{\mu}, V, T) - G(\boldsymbol{\mu}, V, T) = -k_B\left(\frac{\partial \ln Z}{\partial 1/T}\right)_{\boldsymbol{\mu},V} \quad (2.125)$$

$$N_l(\boldsymbol{\mu}, V, T) = \langle N_l' \rangle = k_B T\left(\frac{\partial \ln Z}{\partial \mu_l}\right)_{\mu_l,V,T} \quad (2.126)$$

$$G(\boldsymbol{\mu}, V, T) = \sum_l \mu_l N_l \quad (2.127)$$

$$H(\boldsymbol{\mu}, V, T) = D(\boldsymbol{\mu}, V, T) + G(\boldsymbol{\mu}, V, T) + p(\boldsymbol{\mu}, V, T)V \quad (2.128)$$

$$A(\boldsymbol{\mu}, V, T) = \Phi_G(\boldsymbol{\mu}, V, T) + G(\boldsymbol{\mu}, V, T) \quad (2.129)$$

$$U(\boldsymbol{\mu}, V, T) = \langle U(\boldsymbol{N}', V, T) \rangle \quad (2.130)$$

Equations 2.122–2.130 can then provide (using also their derivatives) any thermodynamic property in terms of the grand-canonical partition function.

Finally, considering as usual that $e^{-\beta[A(\boldsymbol{N}',V,T)-\sum_l \mu_l N_l']}$ is highly peaked with its maximum at $\boldsymbol{N}' = \boldsymbol{N}_M \cong \langle \boldsymbol{N}' \rangle = \boldsymbol{N}$ where

$$\left(\frac{\partial A}{\partial N_l'}\right)_{N_l',V,T} - \mu_l = \mu_l(\boldsymbol{N}', V, T) - \mu_l = 0 \quad (2.131)$$

we have

$$\mu_l(\boldsymbol{N}, V, T) \cong \mu_l \quad (2.132)$$

Therefore, indicating for whatever thermodynamic property X with the notation $X(\boldsymbol{\mu}, V, T)$ the property as obtained within the grand-canonical ensemble and with the notation $X(\boldsymbol{N}, V, T)$ the same property as provided by the canonical ensemble, we obtain

$$p(\boldsymbol{\mu}, V, T) = \langle p(\boldsymbol{N}', V, T) \rangle \cong p(\boldsymbol{N}, V, T) \quad (2.133)$$

$$U(\boldsymbol{\mu}, V, T) = \langle U(\boldsymbol{N}', V, T) \rangle \cong U(\boldsymbol{N}, V, T) \quad (2.134)$$

$$G(\boldsymbol{\mu}, V, T) = \sum_l \mu_l N_l \cong G(\boldsymbol{N}, V, T) \quad (2.135)$$

$$\begin{aligned} \Phi_G(\boldsymbol{\mu}, V, T) &= -p(\boldsymbol{\mu}, V, T)V \cong -p(\boldsymbol{N}, V, T)V \\ &= A(\boldsymbol{N}, V, T) - G(\boldsymbol{N}, V, T) = \Phi_G(\boldsymbol{N}, V, T) \end{aligned} \quad (2.136)$$

$$\begin{aligned} D(\boldsymbol{\mu}, V, T) &= U(\boldsymbol{\mu}, V, T) - G(\boldsymbol{\mu}, V, T) \cong U(\boldsymbol{N}, V, T) - G(\boldsymbol{N}, V, T) \\ &= D(\boldsymbol{N}, V, T) \end{aligned} \quad (2.137)$$

$$\begin{aligned} S(\boldsymbol{\mu}, V, T) &= \frac{D(\boldsymbol{\mu}, V, T) - \Phi_G(\boldsymbol{\mu}, V, T)}{T} \cong \frac{D(\boldsymbol{N}, V, T) - \Phi_G(\boldsymbol{N}, V, T)}{T} \\ &= \frac{U(\boldsymbol{N}, V, T) - A(\boldsymbol{N}, V, T)}{T} = S(\boldsymbol{N}, V, T) \end{aligned} \quad (2.138)$$

the approximations becoming fully accurate for any system equal or larger than one elementary system (i.e., a thermodynamic system). The last equations, together with the results of the previous subsection, then plainly show that the different statistical ensembles provide fully equivalent thermodynamics.

References

1. Landau LD, Lifshitz EM (1969) Statistical physics. Pergamon Press, Oxford
2. Dirac PAM (1958) The principles of quantum mechanics. Clarendon Press, Oxford

Chapter 3
Statistical Mechanics: Application to Chemical Thermodynamics

Abstract The general principles outlined in Chap. 2 and therein specifically employed for a rigorous statistical mechanical description of thermodynamics are further developed in this chapter for treating the central issue of Chemistry: the thermodynamics of the chemical transformations, either in vapor or in condensed phase. After an initial introduction to chemical equilibrium, with specific emphasis on the chemical potential and the related concept of activity, the chapter will illustrate the treatment of the main determinants of whatever problem concerning the chemical equilibrium: the partial molecular properties and the equilibrium constant. In the final part of the chapter two examples are reported: the first one dealing with the calculation of the pKa, i.e., the equilibrium constant for a proton transfer reaction of a Bronsted acid to water in water solution, and the second one illustrating the conformational equilibrium of a simple biomolecule in water solution.

3.1 The Chemical Equilibrium

In Chap. 2 of the book we have described how to obtain a rigorous statistical description of thermodynamics, on the basis of completely general quantum mechanical relations and introducing different statistical ensembles to model real physical processes as occurring at different conditions (i.e., at different fixed state variables). We also showed that although each statistical ensemble provides a different fluctuation behavior, for thermodynamic systems (i.e., equal or larger than an elementary system) any thermodynamic property, as obtained at corresponding thermodynamic conditions, is virtually identical in all the ensembles in line with classical thermodynamics. However, the ensembles used do not explicitly treat chemical reaction processes, requiring to model the interconversion of chemical species.

Let us consider a macroscopic system in either the canonical or the isothermal–isobaric conditions (i.e., no matter exchange with its environment) where we consider the balanced chemical reaction

$$\nu_1 C_1 + \nu_2 C_2 \rightleftharpoons \nu_3 C_3 + \nu_4 C_4 \tag{3.1}$$

A. Amadei and M. Aschi, *Statistical Mechanics for Chemical Thermodynamics and Kinetics*, https://doi.org/10.1007/978-3-031-77929-9_3

with C_1, C_2, C_3, C_4 the four different chemical species involved in the reaction (we assume no other chemical transitions in the system) and $\nu_1, \nu_2, \nu_3, \nu_4$ the corresponding balancing factors (note that the balancing factors are not necessarily integer numbers). We can define the (thermodynamic) chemical equilibrium condition as the system free energy minimum in the space of the reaction chemical species amounts (N_1, N_2, N_3, N_4), when expressing their changes according to the balancing factors

$$dN_2 = \frac{\nu_2}{\nu_1} dN_1 \tag{3.2}$$

$$dN_3 = -\frac{\nu_3}{\nu_1} dN_1 \tag{3.3}$$

$$dN_4 = -\frac{\nu_4}{\nu_1} dN_1 \tag{3.4}$$

Therefore, we can use N_1 as the proper single variable describing the reaction progress and thus the equilibrium is necessarily obtained when in the canonical system we have

$$\begin{aligned} \left(\frac{\partial A}{\partial N_1}\right)_{N_f,V,T} &= \left(\frac{\partial A}{\partial N_1}\right)_{N_1,V,T} + \left(\frac{\partial A}{\partial N_2}\right)_{N_2,V,T} \frac{\nu_2}{\nu_1} \\ &- \left(\frac{\partial A}{\partial N_3}\right)_{N_3,V,T} \frac{\nu_3}{\nu_1} - \left(\frac{\partial A}{\partial N_4}\right)_{N_4,V,T} \frac{\nu_4}{\nu_1} \\ &= \mu_1 + \frac{\nu_2}{\nu_1}\mu_2 - \frac{\nu_3}{\nu_1}\mu_3 - \frac{\nu_4}{\nu_1}\mu_4 = 0 \end{aligned} \tag{3.5}$$

or, similarly, when in the isothermal–isobaric system we find

$$\begin{aligned} \left(\frac{\partial G}{\partial N_1}\right)_{N_f,p,T} &= \left(\frac{\partial G}{\partial N_1}\right)_{N_1,p,T} + \left(\frac{\partial G}{\partial N_2}\right)_{N_2,p,T} \frac{\nu_2}{\nu_1} \\ &- \left(\frac{\partial G}{\partial N_3}\right)_{N_3,p,T} \frac{\nu_3}{\nu_1} - \left(\frac{\partial G}{\partial N_4}\right)_{N_4,p,T} \frac{\nu_4}{\nu_1} \\ &= \mu_1 + \frac{\nu_2}{\nu_1}\mu_2 - \frac{\nu_3}{\nu_1}\mu_3 - \frac{\nu_4}{\nu_1}\mu_4 = 0 \end{aligned} \tag{3.6}$$

where $\boldsymbol{N}_l$ is the vector of all the molecular numbers when removing N_l (see Chap. 2), $\boldsymbol{N}_f$ is the vector of the molecular numbers of the chemical species not involved in the reaction and thus at fixed amounts,

$$\left(\frac{\partial A}{\partial N_1}\right)_{N_f,V,T} \quad \left(\frac{\partial G}{\partial N_1}\right)_{N_f,p,T}$$

correspond to the partial derivatives in N_1 according to the reaction process (i.e., considering N_2, N_3, N_4 functions of N_1) and

$$\left(\frac{\partial A}{\partial N_l}\right)_{N_l,V,T} \quad \left(\frac{\partial G}{\partial N_l}\right)_{N_l,p,T}$$

are the usual partial derivatives in N_l considering all the molecular numbers as independent variables. From Eqs. 3.5 and 3.6 we then readily obtain for the chemical equilibrium condition, regardless of the ensemble used,

$$\nu_3 \, \mu_3 + \nu_4 \, \mu_4 = \nu_1 \, \mu_1 + \nu_2 \, \mu_2 \tag{3.7}$$

which provides the equilibrium chemical composition $(N_1^{eq}, N_2^{eq}, N_3^{eq}, N_4^{eq})$ when assuming a single physically valid solution of Eq. 3.7 corresponding to a free energy minimum. We can further investigate the chemical equilibrium properties by considering a special canonical ensemble, now allowing for the chemical interconversion, including all the chemical compositions due to the reaction and expressing the corresponding Helmholtz free energy via its expansion to second order in N_1 around N_1^{eq}

$$\begin{aligned} A(N_1, \boldsymbol{N}_f, V, T) \cong \; & A(N_1^{eq}, \boldsymbol{N}_f, V, T) + \left(\frac{\partial A}{\partial N_1}\right)_{\boldsymbol{N}_f,V,T} \Delta N_1 + \frac{1}{2}\left(\frac{\partial^2 A}{\partial N_1^2}\right)_{\boldsymbol{N}_f,V,T} \Delta N_1^2 \\ = \; & \frac{1}{2}\left(\frac{\partial^2 A}{\partial N_1^2}\right)_{\boldsymbol{N}_f,V,T} \Delta N_1^2 \end{aligned} \tag{3.8}$$

where $\Delta N_1 = \left(N_1 - N_1^{eq}\right)$, the derivatives of the Helmholtz free energy are clearly obtained at $N_1 = N_1^{eq}$ and

$$\left(\frac{\partial^2 A}{\partial N_1^2}\right)_{\boldsymbol{N}_f,V,T} = \left(\frac{\partial \mu_1}{\partial N_1}\right)_{\boldsymbol{N}_f,V,T} + \frac{\nu_2}{\nu_1}\left(\frac{\partial \mu_2}{\partial N_1}\right)_{\boldsymbol{N}_f,V,T} - \frac{\nu_3}{\nu_1}\left(\frac{\partial \mu_3}{\partial N_1}\right)_{\boldsymbol{N}_f,V,T} - \frac{\nu_4}{\nu_1}\left(\frac{\partial \mu_4}{\partial N_1}\right)_{\boldsymbol{N}_f,V,T} > 0$$

as it is required for a free energy minimum.

From Eq. 3.8, using

$$A_{eq} = A(N_1^{eq}, \boldsymbol{N}_f, V, T) \tag{3.9}$$

$$\sigma_{N_1}^2 = k_B T / \left(\frac{\partial^2 A}{\partial N_1^2}\right)_{\boldsymbol{N}_f,V,T} \tag{3.10}$$

and omitting $\boldsymbol{N}_f$ in the notation, we can then express the system canonical partition function for the chemical equilibrium considered and the corresponding probability distribution for ΔN_1 as

$$\begin{aligned} Q = & \sum_{\Delta N_1} e^{-\beta A(\Delta N_1, V, T)} \cong e^{-\beta A_{eq}} \sum_{\Delta N_1} e^{-\Delta N_1^2/(2\sigma_{N_1}^2)} \\ \cong & \; e^{-\beta A_{eq}} \int_{-\infty}^{\infty} e^{-\Delta N_1^2/(2\sigma_{N_1}^2)} d\Delta N_1 = e^{-\beta A_{eq}} \sqrt{2\pi \sigma_{N_1}^2} \end{aligned} \tag{3.11}$$

$$\rho(\Delta N_1) = \frac{e^{-\beta A(\Delta N_1, V, T)}}{\sum_{\Delta N_1} e^{-\beta A(\Delta N_1, V, T)}} \cong \frac{e^{-\Delta N_1^2/(2\sigma_{N_1}^2)}}{\sqrt{2\pi \sigma_{N_1}^2}} \tag{3.12}$$

readily providing $\langle \Delta N_1^2 \rangle \cong \sigma_{N_1}^2 = k_B T / \left(\partial^2 A / \partial N_1^2 \right)_{N_f, V, T}$. Note that the second-order approximation of the free energy, leading to a Gaussian probability distribution for $\rho(\Delta N_1)$, is accurate as $e^{-\beta A(\Delta N_1, V, T)}$ is highly peaked at $\Delta N_1 = 0$, thus ensuring that for any chemical composition not very close to the equilibrium one the corresponding probability is fully negligible.

The derivations shown clearly indicate the chemical potentials as the determinants of chemical equilibria and corresponding fluctuations of the reaction chemical species. Therefore, in order to model the reaction processes we need a proper statistical mechanical derivation of the chemical potential, in particular for fluid-state systems which are the typical molecular systems of interest in chemistry.

3.2 The Canonical Partition Function of Mixed Quantum–Classical Systems

It is worth to consider the case of a fluid-state macroscopic system where the nuclear coordinates can be typically expressed for all the Hamiltonian eigenstates of interest (i.e., the eigenstates significantly contributing to the partition function) via a set of generalized coordinates $\boldsymbol{q} = \{\boldsymbol{\xi}, \boldsymbol{\beta}\}$ with the $\boldsymbol{\xi}$ subset providing the semiclassical coordinates (typically involving the molecular roto-translational coordinates and the dihedral angles) and the $\boldsymbol{\beta}$ one defined by the quantum degrees of freedom. As described in Chap. 1 such Hamiltonian eigenstates are given by the combination of a wave packet, providing the position in phase space of the semiclassical coordinates and their conjugate momenta, with a vibronic eigenstate (i.e., a given combination of an electron-spin eigenstate with a related vibrational one). The corresponding Hamiltonian eigenvalues can then be expressed (see Chap. 1) as

$$\mathcal{U}_{j_{es}, j_v}(\boldsymbol{\xi}, \boldsymbol{\pi}_\xi) \cong K_\xi(\boldsymbol{\xi}, \boldsymbol{0}, \boldsymbol{\pi}_\xi) + \mathcal{U}_{es, j_{es}}(\boldsymbol{\xi}, \boldsymbol{\beta}_{0, j_{es}}) + \mathcal{U}_{v, j_{es}, j_v}(\boldsymbol{\xi}) \tag{3.13}$$

$$K_\xi(\boldsymbol{\xi}, \boldsymbol{0}, \boldsymbol{\pi}_\xi) = \boldsymbol{\pi}_\xi^T \frac{\widetilde{M}_{\xi,\xi}^{-1}}{2} \boldsymbol{\pi}_\xi \tag{3.14}$$

where j_{es}, j_v are the indices of the electron-spin eigenstate and corresponding vibrational one, respectively, K_ξ is the classical kinetic energy due to the semiclassical coordinates $\boldsymbol{\xi}$ and conjugate momenta $\boldsymbol{\pi}_\xi$ as obtained over the reference $\boldsymbol{\beta} = \boldsymbol{0}$ surface (with $\widetilde{M}_{\xi,\xi}$ the corresponding mass tensor), $\mathcal{U}_{v, j_{es}, j_v}(\boldsymbol{\xi})$ is the vibrational energy (i.e., the j_v vibrational eigenenergy as obtained for the j_{es} electron-spin eigenstate) and $\mathcal{U}_{es, j_{es}}(\boldsymbol{\xi}, \boldsymbol{\beta}_{0, j_{es}})$ is the electron-spin energy over the $\boldsymbol{\beta}_{0, j_{es}}$ surface corresponding to the minimum $\mathcal{U}_{es, j_{es}}$ electron-spin energy at each $\boldsymbol{\xi}$ position. Therefore, replacing in the canonical partition function each Hamiltonian eigenvalue j index (see Chap. 2) with the corresponding combination of the j_{es}, j_v indices and $\boldsymbol{\xi}$, $\boldsymbol{\pi}_\xi$ phase space position, we can rewrite the canonical partition function as

$$Q \cong \sum_{j_{es}} \sum_{j_v} \Theta \int e^{-[\mathcal{H}_{j_{es}}(\boldsymbol{\xi},\boldsymbol{\pi}_\xi)+\mathcal{U}_{v,j_{es},j_v}(\boldsymbol{\xi})]/(k_B T)} \frac{d\boldsymbol{\xi} d\boldsymbol{\pi}_\xi}{h^D} \quad (3.15)$$

$$\mathcal{H}_{j_{es}}(\boldsymbol{\xi},\boldsymbol{\pi}_\xi) = K_\xi(\boldsymbol{\xi},\mathbf{0},\boldsymbol{\pi}_\xi) + \mathcal{U}_{es,j_{es}}(\boldsymbol{\xi},\boldsymbol{\beta}_{0,j_{es}}) \quad (3.16)$$

$$\Theta = \Pi_l (1+\gamma_l)^{-N_l} \, (N_l!)^{-1} \quad (3.17)$$

with $d\boldsymbol{\xi} d\boldsymbol{\pi}_\xi = \Pi_{i=1}^D d\xi_i d\pi_{\xi_i}$, $\mathcal{H}_{j_{es}}(\boldsymbol{\xi},\boldsymbol{\pi}_\xi)$ the classical-like Hamiltonian (i.e., providing the classical equations of motion within the $\boldsymbol{\xi},\boldsymbol{\pi}_\xi$ phase space), l the chemical species index, D the total number of the semiclassical degrees of freedom $\boldsymbol{\xi}$ (h^{-D} being the wave packet density in phase space) and Θ providing the quantum correction for the identical nuclei permutations due to the $\boldsymbol{\xi}$ coordinates (hereafter the molecular roto-translational degrees of freedom and the internal semiclassical coordinates). For each lth chemical species $(1+\gamma_l)^{-N_l}$ provides the correction for the possible identical nuclei permutations within each molecule due to the rotational and (semiclassical) internal coordinates, and $(N_l!)^{-1}$ provides the correction for the permutations of identical molecules. Note that when dealing with systems involving flexible *molecules* with conformational states requiring different *molecular* generalized coordinates, we can obtain the proper partition function simply summing all the partition functions as obtained, by using Eq. 3.15, constraining each flexible *molecule* to be in one of such conformations. Therefore, for sake of simplicity and without loss of generality we will not explicitly treat such an unusual case.

Considering that for any electron-spin eigenstate $\mathcal{U}_{v,j_{es},j_v}(\boldsymbol{\xi})/(k_B T)$ is only slightly changing as a function of the $\boldsymbol{\xi}$ coordinates, defining

$$E_{v,j_{es},j_v} = \frac{\int e^{-[\mathcal{H}_{j_{es}}(\boldsymbol{\xi},\boldsymbol{\pi}_\xi)]/(k_B T)} \, \mathcal{U}_{v,j_{es},j_v}(\boldsymbol{\xi}) \, d\boldsymbol{\xi} d\boldsymbol{\pi}_\xi}{\int e^{-[\mathcal{H}_{j_{es}}(\boldsymbol{\xi},\boldsymbol{\pi}_\xi)]/(k_B T)} \, d\boldsymbol{\xi} d\boldsymbol{\pi}_\xi} \quad (3.18)$$

we can use the first-order expansion of $e^{-\mathcal{U}_{v,j_{es},j_v}/(k_B T)}$ in $\mathcal{U}_{v,j_{es},j_v}$ around E_{v,j_{es},j_v}

$$e^{-\mathcal{U}_{v,j_{es},j_v}(\boldsymbol{\xi})/(k_B T)} \cong e^{-E_{v,j_{es},j_v}/(k_B T)} \left[1 - \frac{\mathcal{U}_{v,j_{es},j_v}(\boldsymbol{\xi}) - E_{v,j_{es},j_v}}{k_B T}\right] \quad (3.19)$$

into Eq. 3.15 to obtain

$$\begin{aligned} Q &\cong \sum_{j_{es}} \sum_{j_v} e^{-E_{v,j_{es},j_v}/(k_B T)} \, \Theta \int e^{-\mathcal{H}_{j_{es}}(\boldsymbol{\xi},\boldsymbol{\pi}_\xi)/(k_B T)} \left[1 - \frac{\mathcal{U}_{v,j_{es},j_v}(\boldsymbol{\xi}) - E_{v,j_{es},j_v}}{k_B T}\right] \frac{d\boldsymbol{\xi} d\boldsymbol{\pi}_\xi}{h^D} \\ &= \sum_{j_{es}} \sum_{j_v} e^{-E_{v,j_{es},j_v}/(k_B T)} \, \Theta \int e^{-\mathcal{H}_{j_{es}}(\boldsymbol{\xi},\boldsymbol{\pi}_\xi)/(k_B T)} \frac{d\boldsymbol{\xi} d\boldsymbol{\pi}_\xi}{h^D} \\ &= \Theta \sum_{j_{es}} Q_{v,j_{es}} \int e^{-\mathcal{H}_{j_{es}}(\boldsymbol{\xi},\boldsymbol{\pi}_\xi)/(k_B T)} \frac{d\boldsymbol{\xi} d\boldsymbol{\pi}_\xi}{h^D} \end{aligned} \quad (3.20)$$

$$Q_{v,j_{es}} = \sum_{j_v} e^{-E_{v,j_{es},j_v}/(k_B T)} \quad (3.21)$$

Finally, realizing that except for very low temperature conditions the nuclear spin contribution to the electron-spin eigenvalues $\mathcal{U}_{es,j_{es}}$ is fully negligible (i.e., negligible compared to k_BT), in the partition function we can use the electronic energies $\mathcal{U}_{e,j_e}$ as provided by the eigenstates of the electron-spin Hamiltonian when disregarding any nuclear spin effects (i.e., the electronic eigenstates typically corresponding to highly separated energy levels). Such nuclear spin effects can be then included in the partition function via a simple degeneration factor N_{spin} virtually identical for all the electronic energies and representing the electron-spin eigenstates with energies virtually identical to each electronic one. Therefore, except for very low or very high temperature conditions, in Eq. 3.20 j_{es} can be substituted by the pure electronic eigenstate index j_e with only the electronic ground eigenstate (i.e., $j_e = 0$) being typically relevant, thus allowing us to write

$$\begin{aligned} Q &\cong N_{spin}\, \Theta \sum_{j_e} Q_{v,j_e} \int e^{-\mathcal{H}_{j_e}(\xi,\pi_\xi)/(k_BT)} \frac{d\xi d\pi_\xi}{h^D} \\ &\cong N_{spin}\, \Theta\, Q_{v,0} \int e^{-\mathcal{H}_0(\xi,\pi_\xi)/(k_BT)} \frac{d\xi d\pi_\xi}{h^D} \end{aligned} \tag{3.22}$$

with $Q_{v,0}$ and $\mathcal{H}_0 = K_\xi + \mathcal{U}_{e,0}$ as obtained fully disregarding the nuclear spin effects. Equation 3.22 provides the usual expression of the canonical partition function utilized to obtain the Helmholtz free energy of molecular systems (in particular liquid–fluid systems)

$$\begin{aligned} A(N,V,T) &\cong -k_BT \ln\Big(N_{spin}\, \Theta\, Q_{v,0}\Big) - k_BT \ln\Big[\int e^{-\mathcal{H}_0(\xi,\pi_\xi)/(k_BT)} \frac{d\xi d\pi_\xi}{h^D}\Big] \\ &= -k_BT \ln\Big(\Gamma\, Q_{v,0}\Big) - k_BT \ln\Big[\int e^{-\mathcal{H}_0(\xi,\pi_\xi)/(k_BT)} \frac{d\xi d\pi_\xi}{h^D}\Big] \end{aligned} \tag{3.23}$$

and hence the complete thermodynamics in terms of the electronic ground eigenstate vibrational partition function $Q_{v,0} \cong \Pi_l (q_{v,0,l})^{N_l}$ and phase space integral ($q_{v,0,l}$ is the lth species electronic ground eigenstate molecular vibrational partition function). It is worth noting that the degeneration factor $N_{spin} \cong \Pi_l (n_l)^{N_l}$ (with n_l the lth species molecular degeneration factor) is often omitted in Eqs. 3.22 and 3.23 as for any kind of temperature or volume change as well as for any (balanced) chemical reaction N_{spin} is invariant, i.e., it only provides a fixed reference entropy value.

3.3 The Chemical Potential

From Eq. 3.23 we can obtain an explicit expression of the chemical potential μ_j for a given chemical species in a fluid-state (macroscopic) system, by using the definition of the chemical potential via the Helmholtz free energy. In fact, omitting in the notation each $N_l \neq N_j$, we have (realizing that a single molecule N_j change is equivalent to a differential, see Chap. 2)

$$\begin{aligned}
\mu_j &= \left(\frac{\partial A}{\partial N_j}\right)_{V,T} = A(N_j+1, V, T) - A(N_j, V, T) \\
&\cong -k_B T \ln\left[\frac{\int e^{-\mathcal{H}_0(N_j)/(k_B T)}\, e^{-[\Delta\mathcal{U}_{e,0}(\boldsymbol{\xi},\boldsymbol{\varrho})+k(\boldsymbol{\varrho},\boldsymbol{\pi}_\varrho)]/(k_B T)}\, d\boldsymbol{\xi} d\boldsymbol{\pi}_\xi d\boldsymbol{\varrho} d\boldsymbol{\pi}_\varrho / h^{d_j}}{\int e^{-\mathcal{H}_0(N_j)/(k_B T)}\, d\boldsymbol{\xi} d\boldsymbol{\pi}_\xi}\right] \\
&- k_B T \ln\left[\frac{n_j\, q_{v,0,j}}{(1+\gamma_j)(N_j+1)}\right] \qquad (3.24)
\end{aligned}$$

where d_j is the number of semiclassical degrees of freedom in a single molecule of the jth chemical species, n_j and $q_{v,0,j}$ are the corresponding molecular nuclear spin degeneration factor and electronic ground eigenstate molecular vibrational partition function (see the previous section), $\mathcal{H}_0(N_j)$ and $\{\boldsymbol{\xi}, \boldsymbol{\pi}_\xi\}$ are the classical-like Hamiltonian and semiclassical coordinates and momenta of the system before including the jth species added molecule,

$$\boldsymbol{\varrho} = \begin{bmatrix} \boldsymbol{R}_G \\ \theta \\ \phi \\ \psi \\ \boldsymbol{\varrho}_{int} \end{bmatrix} \qquad (3.25)$$

are the semiclassical coordinates of the added molecule (i.e., the molecular center of mass Cartesian coordinates $\boldsymbol{R}_G$, the Eulerian angles θ, ϕ, ψ, $0 \leq \theta \leq \pi$, $0 \leq \phi \leq 2\pi$, $0 \leq \psi \leq 2\pi$, and the internal generalized semiclassical coordinates $\boldsymbol{\varrho}_{int}$) and

$$\boldsymbol{\pi}_\varrho = \begin{bmatrix} \boldsymbol{\pi}_{R_G} \\ \pi_\theta \\ \pi_\phi \\ \pi_\psi \\ \boldsymbol{\pi}_{\varrho_{int}} \end{bmatrix} \qquad (3.26)$$

the corresponding conjugate momenta. Moreover,

$$k = \boldsymbol{\pi}_\varrho^T \frac{\widetilde{m}_\varrho^{-1}}{2} \boldsymbol{\pi}_\varrho \qquad (3.27)$$

is the added molecule semiclassical kinetic energy with $\widetilde{m}_\varrho$ the corresponding molecular mass tensor, and thus

$$\Delta\mathcal{U}_{e,0}(\boldsymbol{\xi},\boldsymbol{\varrho}) + k(\boldsymbol{\varrho},\boldsymbol{\pi}_\varrho) = \mathcal{H}_0(N_j+1) - \mathcal{H}_0(N_j) \qquad (3.28)$$

$$\Delta\mathcal{U}_{e,0}(\boldsymbol{\xi},\boldsymbol{\varrho}) = \mathcal{U}_{e,0}(N_j+1) - \mathcal{U}_{e,0}(N_j) \qquad (3.29)$$

where in $\mathcal{U}_{e,0}(N_j+1)$ and $\mathcal{U}_{e,0}(N_j)$ we always have all the nuclear quantum coordinates fixed at their minimum energy position for each semiclassical configuration. By considering that the molecular mass tensor is necessarily a symmetric matrix with hence orthonormal eigenvectors and a unitary Jacobean for the corresponding (orthogonal) momenta transformation, integration over these new momenta $\boldsymbol{\pi}_\lambda$ of the added molecule provides (realizing that the mass tensor determinant is invariant for such a momenta transformation)

$$\int e^{-\mathcal{H}_0(N_j)/(k_BT)}\, e^{-[\Delta\mathcal{U}_{e,0}(\boldsymbol{\xi},\boldsymbol{\varrho})+k(\boldsymbol{\varrho},\boldsymbol{\pi}_\varrho)]/(k_BT)}\, d\boldsymbol{\xi}\, d\boldsymbol{\pi}_\xi\, d\boldsymbol{\varrho}\, d\boldsymbol{\pi}_\lambda$$
$$= \left(2\pi k_BT\right)^{d_j/2}\int e^{-\mathcal{H}_0(N_j)/(k_BT)}\, e^{-\Delta\mathcal{U}_{e,0}(\boldsymbol{\xi},\boldsymbol{\varrho})/(k_BT)}\sqrt{\det\left(\widetilde{m}_\varrho\right)}\, d\boldsymbol{\xi}\, d\boldsymbol{\pi}_\xi\, d\boldsymbol{\varrho}$$

and thus from Eq. 3.24 we have

$$\begin{aligned}\mu_j \cong\; & -k_BT\ln\left[\frac{\int e^{-\mathcal{H}_0(N_j)/(k_BT)}\, e^{-\Delta\mathcal{U}_{e,0}(\boldsymbol{\xi},\boldsymbol{\varrho})/(k_BT)}\sqrt{\det(\widetilde{m}_j)}\sin\theta\, d\boldsymbol{\xi}\, d\boldsymbol{\pi}_\xi\, d\boldsymbol{\varrho}/h^{d_j}}{\int e^{-\mathcal{H}_0(N_j)/(k_BT)}\, d\boldsymbol{\xi}\, d\boldsymbol{\pi}_\xi}\right]\\ & - k_BT\ln\left[\left(2\pi k_BT\right)^{d_j/2}\right] - k_BT\ln\left[\frac{n_j\, q_{v,0,j}}{(1+\gamma_j)(N_j+1)}\right]\end{aligned} \tag{3.30}$$

where $\sin\theta \geq 0$ as $0 \leq \theta \leq \pi$ and $\widetilde{m}_j$ is the added molecule mass tensor for the classical kinetic energy expressed via the angular velocity components within the orthonormal molecular barycentric reference frame (i.e., $\omega_1, \omega_2, \omega_3$) instead via the time derivatives of the Eulerian angles (i.e., $\dot\theta, \dot\phi, \dot\psi$), with $\widetilde{m}_\varrho = \widetilde{T}^T\widetilde{m}_j\widetilde{T}$, the transformation matrix defined by

$$\begin{bmatrix}\dot{\boldsymbol{R}}_G\\ \omega_1\\ \omega_2\\ \omega_3\\ \dot{\boldsymbol{\varrho}}_{int}\end{bmatrix} = \widetilde{T}\begin{bmatrix}\dot{\boldsymbol{R}}_G\\ \dot\theta\\ \dot\phi\\ \dot\psi\\ \dot{\boldsymbol{\varrho}}_{int}\end{bmatrix} \tag{3.31}$$

and hence $\det(\widetilde{m}_\varrho) = \det(\widetilde{T}^T)\det(\widetilde{m}_j)\det(\widetilde{T}) = (\sin\theta)^2\det(\widetilde{m}_j)$ as following from analytical mechanics [1] providing $\det(\widetilde{T}^T) = \det(\widetilde{T}) = \sin\theta$ (note that the Jacobean for the momenta transformation corresponding to the $\dot\theta, \dot\phi, \dot\psi \rightarrow \omega_1, \omega_2, \omega_3$ transformation is [2] $|\sin\theta| = \sin\theta, 0 \leq \theta \leq \pi$).

Furthermore, realizing that for a fluid-state macroscopic system at any fixed roto-translational coordinates $\boldsymbol{\varrho}_{rt} = \{\boldsymbol{R}_G, \theta, \phi, \psi\}$ of the added molecule the integral over all the other coordinates $(\boldsymbol{\xi}, \boldsymbol{\varrho}_{int})$ and momenta $(\boldsymbol{\pi}_\xi)$ is a constant independent of $\boldsymbol{\varrho}_{rt}$, by choosing an arbitrary reference roto-translational position ($\boldsymbol{\varrho}_{rt} = \boldsymbol{\varrho}_{rt}^0$) we can write

$$
\begin{aligned}
\mu_j \cong & -k_B T \ln\Big[\frac{\int e^{-\mathcal{H}_0(N_j)/(k_BT)}\, e^{-\Delta\mathcal{U}^0_{e,0}/(k_BT)}\, \sqrt{\det(\widetilde{m}_j)}\, d\boldsymbol{\xi} d\boldsymbol{\pi}_\xi d\boldsymbol{\varrho}_{int}/h^{d_j}}{\int e^{-\mathcal{H}_0(N_j)/(k_BT)}\, d\boldsymbol{\xi} d\boldsymbol{\pi}_\xi}\Big] \\
& - k_B T \ln\Big[\Big(2\pi k_B T\Big)^{d_j/2}\Big] - k_B T \ln\Big[\frac{n_j\, q_{v,0,j}\, 8\pi^2}{(1+\gamma_j)}\Big] + k_B T \ln\Big[\frac{N_j+1}{V}\Big] \\
\cong & -k_B T \ln\Big[\frac{\int e^{-\mathcal{H}_0(N_j)/(k_BT)}\, e^{-\Delta\mathcal{U}^0_{e,0}/(k_BT)}\, \sqrt{\det(\widetilde{m}_j)}\, d\boldsymbol{\xi} d\boldsymbol{\pi}_\xi d\boldsymbol{\varrho}_{int}/h^{d_j}}{\int e^{-\mathcal{H}_0(N_j)/(k_BT)}\, d\boldsymbol{\xi} d\boldsymbol{\pi}_\xi}\Big] \\
& - k_B T \ln\Big[\Big(2\pi k_B T\Big)^{d_j/2}\Big] - k_B T \ln\Big[\frac{n_j\, q_{v,0,j}\, 8\pi^2}{(1+\gamma_j)}\Big] + k_B T \ln \rho_j
\end{aligned} \tag{3.32}
$$

with $\rho_j = N_j/V \cong (N_j+1)/V$, $\Delta\mathcal{U}^0_{e,0} = \Delta\mathcal{U}_{e,0}(\boldsymbol{\xi}, \boldsymbol{\varrho}_{int}, \boldsymbol{\varrho}^0_{rt})$ and we used $\int d\boldsymbol{R}_G = V$ and $\int \sin\theta\, d\theta d\phi d\psi = 8\pi^2$ (note that for the special case of a rigid linear molecule the integral over the Eulerian angles would be 4π instead of $8\pi^2$). From Eq. 3.32 we then obtain, defining $v_{int,j} = \int d\boldsymbol{\varrho}_{int}$ (the integral being extended only over the accessible $\boldsymbol{\varrho}_{int}$ subspace of the stable added molecule),

$$
\begin{aligned}
\mu_j \cong & -k_B T \ln\Big[\frac{\int e^{-\mathcal{H}_0(N_j)/(k_BT)}\, e^{-\Delta\mathcal{U}^0_{e,0}/(k_BT)}\, \sqrt{\det(\widetilde{m}_j)}\, d\boldsymbol{\xi} d\boldsymbol{\pi}_\xi d\boldsymbol{\varrho}_{int}}{v_{int,j} \int e^{-\mathcal{H}_0(N_j)/(k_BT)}\, d\boldsymbol{\xi} d\boldsymbol{\pi}_\xi}\Big] \\
& - k_B T \ln\Big[\frac{v_{int,j}\, n_j\, q_{v,0,j}\, 8\pi^2}{h^{d_j}(1+\gamma_j)} \Big(2\pi k_B T\Big)^{d_j/2}\Big] + k_B T \ln \rho_j \\
\cong & -k_B T \ln\langle \int e^{-\Delta\mathcal{U}^0_{e,0}/(k_BT)} \sqrt{\det(\widetilde{m}_j)/v^2_{int,j}}\, d\boldsymbol{\varrho}_{int} \rangle_{N_j} \\
& - k_B T \ln\Big[\frac{v_{int,j}\, n_j\, q_{v,0,j}\, 8\pi^2}{h^{d_j}(1+\gamma_j)} \Big(2\pi k_B T\Big)^{d_j/2}\Big] + k_B T \ln \rho_j
\end{aligned} \tag{3.33}
$$

or alternatively (see the first line of the previous equation)

$$
\begin{aligned}
\mu_j \cong & -k_B T \ln\Big[\frac{\int e^{-\mathcal{H}_0(N_j)/(k_BT)}\, e^{-\Delta\mathcal{U}^0_{e,0}/(k_BT)}\, \sqrt{\det(\widetilde{m}_j)}\, d\boldsymbol{\xi} d\boldsymbol{\pi}_\xi d\boldsymbol{\varrho}_{int}}{\int e^{-\mathcal{H}_0(N_j)/(k_BT)}\, e^{[-\Delta\mathcal{U}^0_{e,0}+\Delta\mathcal{U}^0_{e,0}]/(k_BT)} \frac{\sqrt{\det(\widetilde{m}_j)}}{\sqrt{\det(\widetilde{m}_j)}}\, d\boldsymbol{\xi} d\boldsymbol{\pi}_\xi d\boldsymbol{\varrho}_{int}}\Big] \\
& - k_B T \ln\Big[\frac{v_{int,j}\, n_j\, q_{v,0,j}\, 8\pi^2}{h^{d_j}(1+\gamma_j)} \Big(2\pi k_B T\Big)^{d_j/2}\Big] + k_B T \ln \rho_j \\
\cong & \; k_B T \ln\langle e^{\Delta\mathcal{U}^0_{e,0}/(k_BT)} \sqrt{1/\det(\widetilde{m}_j)} \rangle_{N_j+1} \\
& - k_B T \ln\Big[\frac{v_{int,j}\, n_j\, q_{v,0,j}\, 8\pi^2}{h^{d_j}(1+\gamma_j)} \Big(2\pi k_B T\Big)^{d_j/2}\Big] + k_B T \ln \rho_j
\end{aligned} \tag{3.34}
$$

where

$$\begin{aligned}
&\langle \int e^{-\Delta\mathcal{U}^0_{e,0}/(k_BT)}\sqrt{\det(\widetilde{m}_j)/v^2_{int,j}}\, d\boldsymbol{Q}_{int} \rangle_{N_j} \\
&= \int \rho_{N_j}(\boldsymbol{\xi}, \boldsymbol{\pi}_\xi)\Big(\int e^{-\Delta\mathcal{U}^0_{e,0}/(k_BT)}\sqrt{\det(\widetilde{m}_j)/v^2_{int,j}}\, d\boldsymbol{Q}_{int}\Big) d\boldsymbol{\xi} d\boldsymbol{\pi}_\xi \quad (3.35)\\
&\langle e^{\Delta\mathcal{U}^0_{e,0}/(k_BT)}\sqrt{1/\det(\widetilde{m}_j)} \rangle_{N_j+1} \\
&= \int \rho_{N_j+1}(\boldsymbol{\xi}, \boldsymbol{\pi}_\xi, \boldsymbol{Q}_{int})\, e^{\Delta\mathcal{U}^0_{e,0}/(k_BT)}\sqrt{1/\det(\widetilde{m}_j)}\, d\boldsymbol{\xi} d\boldsymbol{\pi}_\xi d\boldsymbol{Q}_{int} \quad (3.36)
\end{aligned}$$

with

$$\rho_{N_j}(\boldsymbol{\xi}, \boldsymbol{\pi}_\xi) = \frac{e^{-\mathcal{H}_0(N_j)/(k_BT)}}{\int e^{-\mathcal{H}_0(N_j)/(k_BT)}\, d\boldsymbol{\xi} d\boldsymbol{\pi}_\xi} \quad (3.37)$$

$$\rho_{N_j+1}(\boldsymbol{\xi}, \boldsymbol{\pi}_\xi, \boldsymbol{Q}_{int}) = \frac{e^{-[\mathcal{H}_0(N_j)+\Delta\mathcal{U}^0_{e,0}]/(k_BT)}\sqrt{\det(\widetilde{m}_j)}}{\int e^{-[\mathcal{H}_0(N_j)+\Delta\mathcal{U}^0_{e,0}]/(k_BT)}\sqrt{\det(\widetilde{m}_j)}\, d\boldsymbol{\xi} d\boldsymbol{\pi}_\xi d\boldsymbol{Q}_{int}} \quad (3.38)$$

the probability densities in the $\{\boldsymbol{\xi}, \boldsymbol{\pi}_\xi\}$ phase space and within the $\{\boldsymbol{\xi}, \boldsymbol{Q}_{int}, \boldsymbol{\pi}_\xi\}$ space, providing averaging either in the N_j or $N_j + 1$ ensemble, respectively, as indicated by the N_j or $N_j + 1$ subscript of the angle brackets. Note that

$$\langle \int e^{-\Delta\mathcal{U}^0_{e,0}/(k_BT)}\sqrt{\det(\widetilde{m}_j)/v^2_{int,j}}\, d\boldsymbol{Q}_{int} \rangle_{N_j} = \langle e^{-\mathcal{F}_{N_j}/(k_BT)}\rangle_{N_j} \quad (3.39)$$

$$\langle e^{\Delta\mathcal{U}^0_{e,0}/(k_BT)}\sqrt{1/\det(\widetilde{m}_j)} \rangle_{N_j+1} = \langle e^{\mathcal{F}_{N_j+1}/(k_BT)}\rangle_{N_j+1} \quad (3.40)$$

are the moment generating functions (MGFs) for the fluctuation distributions of $\mathcal{F}_{N_j}(\boldsymbol{\xi}) = -k_BT\ln\Big[\int e^{-\Delta\mathcal{U}^0_{e,0}/(k_BT)}\sqrt{\det(\widetilde{m}_j)/v^2_{int,j}}\, d\boldsymbol{Q}_{int}\Big]$ and $\mathcal{F}_{N_j+1}(\boldsymbol{\xi}, \boldsymbol{Q}_{int}) = \Delta\mathcal{U}^0_{e,0} - \frac{k_BT}{2}\ln\det(\widetilde{m}_j)$, in the N_j and $N_j + 1$ ensemble, respectively. Moreover, from Eqs. 3.33 and 3.34 we obviously have

$$1/\langle e^{\Delta\mathcal{U}^0_{e,0}/(k_BT)}\sqrt{1/\det(\widetilde{m}_j)} \rangle_{N_j+1} = \langle \int e^{-\Delta\mathcal{U}^0_{e,0}/(k_BT)}\sqrt{\det(\widetilde{m}_j)/v^2_{int,j}}\, d\boldsymbol{Q}_{int} \rangle_{N_j}$$

that is $1/\langle e^{\mathcal{F}_{N_j+1}/(k_BT)}\rangle_{N_j+1} = \langle e^{-\mathcal{F}_{N_j}/(k_BT)}\rangle_{N_j}$.

It is worth noting that when the jth species has no internal semiclassical coordinates $\boldsymbol{Q}_{int}$, then in the previous equations the integration over such coordinates and $v_{int,j}$ disappear.

Finally, let us consider two (A and B) chemical species with their chemical potential difference

$$\begin{aligned}
\mu_B - \mu_A &= \Big[A(N_A, N_B+1) - A(N_A, N_B)\Big] - \Big[A(N_A+1, N_B) - A(N_A, N_B)\Big] \\
&= A(N_A, N_B+1) - A(N_A+1, N_B) \quad (3.41)
\end{aligned}$$

as obtained by using Eq. 3.32 (we consider the general case of non-linear molecules)

$$
\begin{aligned}
\Delta\mu = \ & \mu_B - \mu_A \cong -k_B T \ln\Big[\frac{\int e^{-\mathcal{U}_B/(k_BT)}\ \sqrt{\det(\widetilde{m}_B)}\ d\boldsymbol{\xi} d\boldsymbol{\pi}_\xi d\boldsymbol{Q}^B_{int}/h^{d_B}}{\int e^{-\mathcal{U}_A/(k_BT)}\ \sqrt{\det(\widetilde{m}_A)}\ d\boldsymbol{\xi} d\boldsymbol{\pi}_\xi d\boldsymbol{Q}^A_{int}/h^{d_A}}\Big] \\
- \ & k_B T \ln\Big[\frac{n_B\ q_{v,0,B}(1+\gamma_A)}{n_A\ q_{v,0,A}(1+\gamma_B)}\Big] - \frac{(d_B - d_A)k_B T}{2} \ln(2\pi k_B T) + k_B T \ln\Big[\frac{\rho_B}{\rho_A}\Big] \\
= \ & -k_B T \ln\Big[\frac{\int e^{-\mathcal{U}_A/(k_BT)} \Big(\int e^{-\Delta\mathcal{U}/(k_BT)}\ \sqrt{\frac{\det(\widetilde{m}_B)}{\det(\widetilde{m}_A)}}\ d\boldsymbol{Q}^B_{int}/h^{d_B}\Big)\sqrt{\det(\widetilde{m}_A)} d\boldsymbol{\xi} d\boldsymbol{\pi}_\xi d\boldsymbol{Q}^A_{int}}{v_{int,A}\ \int e^{-\mathcal{U}_A/(k_BT)}\ \sqrt{\det(\widetilde{m}_A)}\ d\boldsymbol{\xi} d\boldsymbol{\pi}_\xi d\boldsymbol{Q}^A_{int}/h^{d_A}}\Big] \\
- \ & k_B T \ln\Big[\frac{n_B\ q_{v,0,B}(1+\gamma_A)}{n_A\ q_{v,0,A}(1+\gamma_B)}\Big] - \frac{(d_B - d_A)k_B T}{2} \ln(2\pi k_B T) + k_B T \ln\Big[\frac{\rho_B}{\rho_A}\Big] \\
= \ & -k_B T \ln\langle \int e^{-\Delta\mathcal{U}/(k_BT)}\ \sqrt{\frac{\det(\widetilde{m}_B)}{\det(\widetilde{m}_A)}}\ d\boldsymbol{Q}^B_{int}\rangle_{N_A+1} + k_B T \ln\Big[v_{int,A} h^{d_B}/h^{d_A}\Big] \\
- \ & k_B T \ln\Big[\frac{n_B\ q_{v,0,B}(1+\gamma_A)}{n_A\ q_{v,0,A}(1+\gamma_B)}\Big] - \frac{(d_B - d_A)k_B T}{2} \ln(2\pi k_B T) + k_B T \ln\Big[\frac{\rho_B}{\rho_A}\Big] \\
= \ & -k_B T \ln\langle \int e^{-[\mathcal{F}_{N_B+1} - \mathcal{F}_{N_A+1}]/(k_BT)}\ d\boldsymbol{Q}^B_{int}\rangle_{N_A+1} + k_B T \ln\Big[v_{int,A} h^{d_B}/h^{d_A}\Big] \\
- \ & k_B T \ln\Big[\frac{n_B\ q_{v,0,B}(1+\gamma_A)}{n_A\ q_{v,0,A}(1+\gamma_B)}\Big] - \frac{(d_B - d_A)k_B T}{2} \ln(2\pi k_B T) + k_B T \ln\Big[\frac{\rho_B}{\rho_A}\Big]
\end{aligned} \tag{3.42}
$$

where $\boldsymbol{Q}^A_{int}$, $\boldsymbol{Q}^B_{int}$ are the A and B added molecule semiclassical internal coordinates, $\Delta\mathcal{U} = \mathcal{U}_B - \mathcal{U}_A$ and

$$\mathcal{U}_B = \mathcal{H}_0(N_A, N_B) + \Delta\mathcal{U}^0_{e,0,B} \tag{3.43}$$

$$\mathcal{U}_A = \mathcal{H}_0(N_A, N_B) + \Delta\mathcal{U}^0_{e,0,A} \tag{3.44}$$

$$\mathcal{F}_{N_B+1} - \mathcal{F}_{N_A+1} = \Delta\mathcal{U} - \frac{k_B T}{2} \ln\Big[\det(\widetilde{m}_B)/\det(\widetilde{m}_A)\Big] \tag{3.45}$$

with $\Delta\mathcal{U}^0_{e,0,A}$, $\Delta\mathcal{U}^0_{e,0,B}$ the electronic ground eigenstate energy shift due to the added A or B molecule at the same fixed roto-translational position, respectively.

3.4 Standard State and Activity Coefficient

3.4.1 *The Vapor–Gas Case*

When considering the vapor–gas condition the standard state MGF simply corresponds to the one as obtained for the pure ideal gas behavior, with hence $\Delta\mathcal{U}^0_{e,0}$ excluding any intermolecular interaction. From Eqs. 3.33 and 3.34 we then obtain

$$\begin{aligned}
\mu_j &\cong -k_BT\ln\langle e^{-\mathcal{F}_{N_j}/(k_BT)}\rangle^{\ominus}_{N_j} - k_BT\ln\Big[\frac{v_{int,j}\; n_j\; q_{v,0,j}\; 8\pi^2}{h^{d_j}(1+\gamma_j)\rho^{\ominus}}\Big(2\pi k_BT\Big)^{d_j/2}\Big] \\
&+ k_BT\ln\Big(\alpha_j\frac{\rho_j}{\rho^{\ominus}}\Big) \\
&= k_BT\ln\langle e^{\mathcal{F}_{N_j+1}/(k_BT)}\rangle^{\ominus}_{N_j+1} - k_BT\ln\Big[\frac{v_{int,j}\; n_j\; q_{v,0,j}\; 8\pi^2}{h^{d_j}(1+\gamma_j)\rho^{\ominus}}\Big(2\pi k_BT\Big)^{d_j/2}\Big] \\
&+ k_BT\ln\Big(\alpha_j\frac{\rho_j}{\rho^{\ominus}}\Big)
\end{aligned} \tag{3.46}$$

$$\begin{aligned}
\alpha_j &= \frac{\langle \int e^{-\Delta\mathcal{U}^0_{e,0}/(k_BT)}\sqrt{\det(\widetilde{m}_j)/v^2_{int,j}}\; d\mathbf{Q}_{int}\,\rangle^{\ominus}_{N_j}}{\langle \int e^{-\Delta\mathcal{U}^0_{e,0}/(k_BT)}\sqrt{\det(\widetilde{m}_j)/v^2_{int,j}}\; d\mathbf{Q}_{int}\,\rangle_{N_j}} = \frac{\langle e^{-\mathcal{F}_{N_j}/(k_BT)}\rangle^{\ominus}_{N_j}}{\langle e^{-\mathcal{F}_{N_j}/(k_BT)}\rangle_{N_j}} \\
&= \frac{\langle e^{\Delta\mathcal{U}^0_{e,0}/(k_BT)}\sqrt{1/\det(\widetilde{m}_j)}\,\rangle_{N_j+1}}{\langle e^{\Delta\mathcal{U}^0_{e,0}/(k_BT)}\sqrt{1/\det(\widetilde{m}_j)}\,\rangle^{\ominus}_{N_j+1}} = \frac{\langle e^{\mathcal{F}_{N_j+1}/(k_BT)}\rangle_{N_j+1}}{\langle e^{\mathcal{F}_{N_j+1}/(k_BT)}\rangle^{\ominus}_{N_j+1}}
\end{aligned} \tag{3.47}$$

where $\rho^{\ominus} = p^{\ominus}/(k_BT)$ is the vapor–gas standard state density as defined as the ideal gas density at the same temperature of the actual system and at the standard pressure $p^{\ominus} = 1$ atm (i.e., the standard state is defined at the standard pressure with however the ideal gas behavior). Considering $q_{v,0,j}$ as independent of the chemical species densities (an accurate approximation for vapor–gas conditions), α_j provides the activity coefficient of the jth species and thus, defining the ideal gas volume of the system by $V_{id} = k_BT\sum_j N_j/p$ with p the equilibrium pressure of the actual system, we have

$$\begin{aligned}
\mu_j &\cong \mu^{\ominus}_j + k_BT\ln\Big(\alpha_j\frac{\rho_j}{\rho^{\ominus}}\Big) = \mu^{\ominus}_j + k_BT\ln\Big(\alpha_j\frac{k_BTN_j}{V_{id}}\frac{V_{id}}{Vp^{\ominus}}\Big) \\
&= \mu^{\ominus}_j + k_BT\ln\Big(\alpha_j\frac{pN_j}{\sum_j N_j}\frac{V_{id}}{Vp^{\ominus}}\Big) = \mu^{\ominus}_j + k_BT\ln\Big(\alpha_j\frac{p\chi_jV_{id}/V}{p^{\ominus}}\Big) \\
&= \mu^{\ominus}_j + k_BT\ln\Big(\alpha_j\frac{\chi_jp_{id}}{p^{\ominus}}\Big) = \mu^{\ominus}_j + k_BT\ln\Big(\alpha_j\frac{p_j}{p^{\ominus}}\Big) \\
&= \mu^{\ominus}_j + k_BT\ln\Big(\frac{f_j}{p^{\ominus}}\Big)
\end{aligned} \tag{3.48}$$

with V the actual vapor–gas system volume, $p_{id} = pV_{id}/V = k_BT\sum_j N_j/V$ the system ideal gas pressure, $\chi_j = N_j/\sum_j N_j$ the jth species molecular fraction, $p_j = \chi_jp_{id} = N_jk_BT/V$ the ideal gas jth species partial pressure, $f_j = \alpha_jp\chi_jV_{id}/V = \alpha_j\chi_jp_{id} = \alpha_jp_j$ the jth species fugacity and

$$\begin{aligned}
\mu_j^{\ominus} &\cong -k_B T \ln \langle e^{-\mathcal{F}_{N_j}/(k_B T)} \rangle^{\ominus}_{N_j} \\
&- k_B T \ln \Big[\frac{v_{int,j}\; n_j\; q_{v,0,j}\; 8\pi^2 k_B T}{h^{d_j}(1+\gamma_j)\; p^{\ominus}} \Big(2\pi k_B T\Big)^{d_j/2} \Big] \\
&= k_B T \ln \langle e^{\mathcal{F}_{N_j+1}/(k_B T)} \rangle^{\ominus}_{N_j+1} \\
&- k_B T \ln \Big[\frac{v_{int,j}\; n_j\; q_{v,0,j}\; 8\pi^2 k_B T}{h^{d_j}(1+\gamma_j)\; p^{\ominus}} \Big(2\pi k_B T\Big)^{d_j/2} \Big]
\end{aligned} \tag{3.49}$$

the (vapor–gas) jth species standard state chemical potential.

Finally, note that for $p \to 0$ we necessarily have $V \to V_{id}$, $\alpha_j \to 1$ and thus $f_j \to \chi_j p_{id} = p_j$.

3.4.2 *The Solute Case*

When considering a solution with the jth species corresponding to a solute component we can define the standard state MGFs via

$$\begin{aligned}
&\lim_{\boldsymbol{\rho}_{solute}\to \mathbf{0}} \langle \int e^{-\Delta \mathcal{U}^0_{e,0}/(k_B T)} \sqrt{\det(\widetilde{m}_j)/v^2_{int,j}}\; d\boldsymbol{Q}_{int} \rangle_{N_j} \\
&= \langle \int e^{-\Delta \mathcal{U}^0_{e,0}/(k_B T)} \sqrt{\det(\widetilde{m}_j)/v^2_{int,j}}\; d\boldsymbol{Q}_{int} \rangle^{\ominus}_{N_j} = \langle e^{-\mathcal{F}_{N_j}/(k_B T)} \rangle^{\ominus}_{N_j}
\end{aligned} \tag{3.50}$$

$$\begin{aligned}
&\lim_{\boldsymbol{\rho}_{solute}\to \mathbf{0}} \langle e^{\Delta \mathcal{U}^0_{e,0}/(k_B T)} \sqrt{1/\det(\widetilde{m}_j)} \rangle_{N_j+1} \\
&= \langle e^{\Delta \mathcal{U}^0_{e,0}/(k_B T)} \sqrt{1/\det(\widetilde{m}_j)} \rangle^{\ominus}_{N_j+1} = \langle e^{\mathcal{F}_{N_j+1}/(k_B T)} \rangle^{\ominus}_{N_j+1}
\end{aligned} \tag{3.51}$$

with $\boldsymbol{\rho}_{solute}$ the vector of all the solute densities and the $\boldsymbol{\rho}_{solute} \to \mathbf{0}$ limit of the expectation value in the canonical ensemble as obtained either by adding solvent molecules or by removing solute molecules at fixed pressure and temperature. Therefore, from Eqs. 3.33 and 3.34 we obtain

$$\begin{aligned}
\mu_j &\cong -k_B T \ln \langle e^{-\mathcal{F}_{N_j}/(k_B T)} \rangle^{\ominus}_{N_j} - k_B T \ln \Big[\frac{v_{int,j}\; n_j\; q_{v,0,j}\; 8\pi^2}{h^{d_j}(1+\gamma_j)\rho^{\ominus}} \Big(2\pi k_B T\Big)^{d_j/2} \Big] \\
&+ k_B T \ln \Big(\alpha_j \frac{\rho_j}{\rho^{\ominus}}\Big) \\
&= k_B T \ln \langle e^{\mathcal{F}_{N_j+1}/(k_B T)} \rangle^{\ominus}_{N_j+1} - k_B T \ln \Big[\frac{v_{int,j}\; n_j\; q_{v,0,j}\; 8\pi^2}{h^{d_j}(1+\gamma_j)\rho^{\ominus}} \Big(2\pi k_B T\Big)^{d_j/2} \Big] \\
&+ k_B T \ln \Big(\alpha_j \frac{\rho_j}{\rho^{\ominus}}\Big)
\end{aligned} \tag{3.52}$$

$$\alpha_j = \frac{\langle \int e^{-\Delta\mathcal{U}^0_{e,0}/(k_B T)} \sqrt{\det(\widetilde{m}_j)/v^2_{int,j}}\, d\boldsymbol{Q}_{int} \rangle^{\ominus}_{N_j}}{\langle \int e^{-\Delta\mathcal{U}^0_{e,0}/(k_B T)} \sqrt{\det(\widetilde{m}_j)/v^2_{int,j}}\, d\boldsymbol{Q}_{int} \rangle_{N_j}} = \frac{\langle e^{-\mathcal{F}_{N_j}/(k_B T)} \rangle^{\ominus}_{N_j}}{\langle e^{-\mathcal{F}_{N_j}/(k_B T)} \rangle_{N_j}}$$

$$= \frac{\langle e^{\Delta\mathcal{U}^0_{e,0}/(k_B T)} \sqrt{1/\det(\widetilde{m}_j)} \rangle_{N_j+1}}{\langle e^{\Delta\mathcal{U}^0_{e,0}/(k_B T)} \sqrt{1/\det(\widetilde{m}_j)} \rangle^{\ominus}_{N_j+1}} = \frac{\langle e^{\mathcal{F}_{N_j+1}/(k_B T)} \rangle_{N_j+1}}{\langle e^{\mathcal{F}_{N_j+1}/(k_B T)} \rangle^{\ominus}_{N_j+1}} \tag{3.53}$$

where $\rho^{\ominus}$ is the standard state solute density, independent of temperature, pressure and solute type and typically corresponding to 1 M (it is worth to remark that for a solute species the standard state is defined at the standard density with however the solute infinite dilution behavior). Considering $q_{v,0,j}$ as virtually independent of the solute densities (at least for not too high densities), α_j provides the activity coefficient of the jth solute species and thus

$$\mu_j \cong \mu^{\ominus}_j + k_B T \ln\left(\alpha_j \frac{\rho_j}{\rho^{\ominus}}\right) = \mu^{\ominus}_j + k_B T \ln\left(\frac{a_j}{\rho^{\ominus}}\right) \tag{3.54}$$

with $a_j = \alpha_j \rho_j$ the jth solute species activity and

$$\begin{aligned}
\mu^{\ominus}_j \cong & -k_B T \ln\langle e^{-\mathcal{F}_{N_j}/(k_B T)} \rangle^{\ominus}_{N_j} \\
& - k_B T \ln\left[\frac{v_{int,j}\, n_j\, q_{v,0,j}\, 8\pi^2}{h^{d_j}(1+\gamma_j)\rho^{\ominus}} \left(2\pi k_B T\right)^{d_j/2}\right] \\
= & \; k_B T \ln\langle e^{\mathcal{F}_{N_j+1}/(k_B T)} \rangle^{\ominus}_{N_j+1} \\
& - k_B T \ln\left[\frac{v_{int,j}\, n_j\, q_{v,0,j}\, 8\pi^2}{h^{d_j}(1+\gamma_j)\rho^{\ominus}} \left(2\pi k_B T\right)^{d_j/2}\right]
\end{aligned} \tag{3.55}$$

the jth solute species standard state chemical potential.

Finally, note that if for all solute species $\rho_j \rightarrow 0$, we necessarily have $\alpha_j \rightarrow 1$ and thus $a_j \rightarrow \rho_j$.

3.4.3 *The Solvent Case*

When considering the jth chemical species as corresponding to the solvent, we can use the previous equations simply setting the standard state identical to the pure solvent condition, e.g., $\rho^{\ominus} = \rho^*$ with ρ^* the pure solvent density (i.e., $\boldsymbol{\rho}_{solute} = \mathbf{0}$) at the same temperature and pressure of the actual condition considered (i.e., ρ^* is a function of temperature, pressure and solvent type). Therefore, assuming the solvent molecular vibrational partition function as independent of the solute density, the solvent chemical potential can be obtained via

$$
\begin{aligned}
\mu_s &\cong \mu^* + k_B T \ln\left(\alpha_s \frac{\rho_s}{\rho^*}\right) \\
&\cong -k_B T \ln\langle e^{-\mathcal{F}_{N_s}/(k_B T)}\rangle^*_{N_s} - k_B T \ln\left[\frac{v_{int,s}\, n_s\, q_{v,0,s}\, 8\pi^2}{h^{d_s}(1+\gamma_s)\rho^*}\left(2\pi k_B T\right)^{d_s/2}\right] \\
&+ k_B T \ln\left(\alpha_s \frac{\rho_s}{\rho^*}\right) \\
&= k_B T \ln\langle e^{\mathcal{F}_{N_s+1}/(k_B T)}\rangle^*_{N_s+1} - k_B T \ln\left[\frac{v_{int,s}\, n_s\, q_{v,0,s}\, 8\pi^2}{h^{d_s}(1+\gamma_s)\rho^*}\left(2\pi k_B T\right)^{d_s/2}\right] \\
&+ k_B T \ln\left(\alpha_s \frac{\rho_s}{\rho^*}\right)
\end{aligned} \tag{3.56}
$$

$$
\begin{aligned}
\alpha_s &= \frac{\langle \int e^{-\Delta\mathcal{U}^0_{e,0}/(k_B T)} \sqrt{\det(\widetilde{m}_s)/v^2_{int,s}}\; d\boldsymbol{Q}_{int}\,\rangle^*_{N_s}}{\langle \int e^{-\Delta\mathcal{U}^0_{e,0}/(k_B T)} \sqrt{\det(\widetilde{m}_s)/v^2_{int,s}}\; d\boldsymbol{Q}_{int}\,\rangle_{N_s}} = \frac{\langle e^{-\mathcal{F}_{N_s}/(k_B T)}\rangle^*_{N_s}}{\langle e^{-\mathcal{F}_{N_s}/(k_B T)}\rangle_{N_s}} \\
&= \frac{\langle e^{\Delta\mathcal{U}^0_{e,0}/(k_B T)} \sqrt{1/\det(\widetilde{m}_s)}\,\rangle_{N_s+1}}{\langle e^{\Delta\mathcal{U}^0_{e,0}/(k_B T)} \sqrt{1/\det(\widetilde{m}_s)}\,\rangle^*_{N_s+1}} = \frac{\langle e^{\mathcal{F}_{N_s+1}/(k_B T)}\rangle_{N_s+1}}{\langle e^{\mathcal{F}_{N_s+1}/(k_B T)}\rangle^*_{N_s+1}}
\end{aligned} \tag{3.57}
$$

where in these equations the s subscript refers to the solvent and the star superscript indicates the solvent standard state (i.e., pure solvent condition). It is worth noting that at high solute dilution (i.e., in the solution the solvent partial molecular properties are virtually identical to the pure solvent ones and $\alpha_s \cong 1$) we can express ρ_s/ρ^* via

$$
\frac{\rho_s}{\rho^*} = \frac{N_s/V}{N_s/V^*} \cong \frac{N_s\, V^*_m}{N_s\, V^*_m + N_{solute}\, V_m} \tag{3.58}
$$

with N_s the number of solvent molecules, V^*_m the partial molecular volume of the pure solvent, $V^* = N_s V^*_m$ the pure solvent total volume (at the same temperature and pressure) and $V_m = \sum_j N_j V_{m,j}/N_{solute}$ the average solute partial molecular volume with j running over the solute species, N_j, $V_{m,j}$ the corresponding molecular number and solute partial molecular volume and $N_{solute} = \sum_j N_j$ the total number of solute molecules. If we consider solute molecules with a mean partial molecular volume similar to the solvent one, i.e., $V_m \approx V^*_m$ and hence

$$
\frac{\rho_s}{\rho^*} \approx \frac{N_s}{N_s + N_{solute}} = \chi_s \tag{3.59}
$$

we readily obtain

$$
\mu_s \approx \mu^* + k_B T \ln(\chi_s) \tag{3.60}
$$

indicating that for any solute–solvent mixture it is worth to use the general expression (see Eq. 3.56)

$$\mu_s \cong \mu^* + k_B T \ln(\alpha_s' \chi_s) = \mu^* + k_B T \ln(a_s) \tag{3.61}$$

$$\alpha_s' = \alpha_s \frac{\rho_s/\rho^*}{\chi_s} \tag{3.62}$$

with a_s the solvent activity.

3.5 The Partial Molecular Properties as Derivatives of the Chemical Potential

From the expressions derived for the chemical potential, it is easy to obtain any partial molecular property using basic thermodynamic relations. In fact for the lth species partial molecular entropy $S_{m,l}$, volume $V_{m,l}$ and enthalpy $H_{m,l}$ we can write

$$\begin{aligned} S_{m,l} &= \left(\frac{\partial S}{\partial N_l}\right)_{N_l,p,T} = \left(\frac{\partial}{\partial N_l}\left(-\frac{\partial G}{\partial T}\right)_{N,p}\right)_{N_l,p,T} \\ &= -\left(\frac{\partial}{\partial T}\left(\frac{\partial G}{\partial N_l}\right)_{N_l,p,T}\right)_{N,p} = -\left(\frac{\partial \mu_l}{\partial T}\right)_{N,p} \end{aligned} \tag{3.63}$$

$$\begin{aligned} V_{m,l} &= \left(\frac{\partial V}{\partial N_l}\right)_{N_l,p,T} = \left(\frac{\partial}{\partial N_l}\left(\frac{\partial G}{\partial p}\right)_{N,T}\right)_{N_l,p,T} \\ &= \left(\frac{\partial}{\partial p}\left(\frac{\partial G}{\partial N_l}\right)_{N_l,p,T}\right)_{N,T} = \left(\frac{\partial \mu_l}{\partial p}\right)_{N,T} \end{aligned} \tag{3.64}$$

$$\begin{aligned} H_{m,l} &= \left(\frac{\partial H}{\partial N_l}\right)_{N_l,p,T} = \left(\frac{\partial}{\partial N_l}\left(\frac{\partial (G/T)}{\partial 1/T}\right)_{N,p}\right)_{N_l,p,T} \\ &= \left(\frac{\partial}{\partial 1/T}\left(\frac{\partial (G/T)}{\partial N_l}\right)_{N_l,p,T}\right)_{N,p} = \left(\frac{\partial (\mu_l/T)}{\partial 1/T}\right)_{N,p} \end{aligned} \tag{3.65}$$

All the other partial molecular properties can then be obtained from the partial molecular entropy, volume and enthalpy via their derivatives in p or T; e.g., the (isobaric) partial molecular heat capacity $c_{p,l}$ is

$$\begin{aligned} c_{p,l} &= \left(\frac{\partial C_p}{\partial N_l}\right)_{N_l,p,T} = \left(\frac{\partial}{\partial N_l}\left(\frac{\partial H}{\partial T}\right)_{N,p}\right)_{N_l,p,T} \\ &= \left(\frac{\partial}{\partial T}\left(\frac{\partial H}{\partial N_l}\right)_{N_l,p,T}\right)_{N,p} = \left(\frac{\partial H_{m,l}}{\partial T}\right)_{N,p} \end{aligned} \tag{3.66}$$

with $C_p = (\partial H/\partial T)_{N,p}$ the isobaric heat capacity of the whole system.

3.6 The Equilibrium Constant

Let us consider the chemical reaction of Eq. 3.1

$$\nu_1 C_1 + \nu_2 C_2 \rightleftharpoons \nu_3 C_3 + \nu_4 C_4$$

in either the vapor–gas or the solution condition with thus the chemical potentials of the reaction species (gas components or solutes) given by Eqs. 3.48 or 3.54. The chemical equilibrium condition is then obtained, according to Eq. 3.7, by

$$\nu_3\mu_3^\ominus + \nu_4\mu_4^\ominus - \nu_1\mu_1^\ominus - \nu_2\mu_2^\ominus = -k_B T \ln \frac{(f_3/p^\ominus)^{\nu_3}(f_4/p^\ominus)^{\nu_4}}{(f_1/p^\ominus)^{\nu_1}(f_2/p^\ominus)^{\nu_2}} \tag{3.67}$$

for the vapor–gas state system and by

$$\nu_3\mu_3^\ominus + \nu_4\mu_4^\ominus - \nu_1\mu_1^\ominus - \nu_2\mu_2^\ominus = -k_B T \ln \frac{(a_3/\rho^\ominus)^{\nu_3}(a_4/\rho^\ominus)^{\nu_4}}{(a_1/\rho^\ominus)^{\nu_1}(a_2/\rho^\ominus)^{\nu_2}} \tag{3.68}$$

for the solutes reaction. From the last equations, defining with

$$\Delta\mu^\ominus = \nu_3\mu_3^\ominus + \nu_4\mu_4^\ominus - \nu_1\mu_1^\ominus - \nu_2\mu_2^\ominus \tag{3.69}$$

the reaction standard state free energy, we readily obtain for the vapor–gas state system

$$e^{-\beta\Delta\mu^\ominus} = \frac{\alpha_3^{\nu_3}\alpha_4^{\nu_4}}{\alpha_1^{\nu_1}\alpha_2^{\nu_2}} \frac{(p_3/p^\ominus)^{\nu_3}(p_4/p^\ominus)^{\nu_4}}{(p_1/p^\ominus)^{\nu_1}(p_2/p^\ominus)^{\nu_2}} = K_{eq} \tag{3.70}$$

and for the solutes reaction

$$e^{-\beta\Delta\mu^\ominus} = \frac{\alpha_3^{\nu_3}\alpha_4^{\nu_4}}{\alpha_1^{\nu_1}\alpha_2^{\nu_2}} \frac{(\rho_3/\rho^\ominus)^{\nu_3}(\rho_4/\rho^\ominus)^{\nu_4}}{(\rho_1/\rho^\ominus)^{\nu_1}(\rho_2/\rho^\ominus)^{\nu_2}} = K_{eq} \tag{3.71}$$

where K_{eq} is the equilibrium constant of the reaction fully defined, at each temperature and pressure of the system, by $\Delta\mu^\ominus$ and α_j, p_j, ρ_j are obtained at the equilibrium conditions, i.e., $p_j = N_j^{eq} k_B T/V$ and $\rho_j = N_j^{eq}/V$. It is worth noting that for typical reactant–product densities the activity coefficients of the reaction species are close to unity (i.e., $\alpha_1 \cong \alpha_2 \cong \alpha_3 \cong \alpha_4 \cong 1$), and thus Eqs. 3.70 and 3.71 reduce to

$$K_{eq} \cong \frac{(p_3/p^\ominus)^{\nu_3}(p_4/p^\ominus)^{\nu_4}}{(p_1/p^\ominus)^{\nu_1}(p_2/p^\ominus)^{\nu_2}} \tag{3.72}$$

and

$$K_{eq} \cong \frac{(\rho_3/\rho^\ominus)^{\nu_3}(\rho_4/\rho^\ominus)^{\nu_4}}{(\rho_1/\rho^\ominus)^{\nu_1}(\rho_2/\rho^\ominus)^{\nu_2}} \tag{3.73}$$

When considering in Eq. 3.1 either $\nu_4 = 0$ or $\nu_2 = \nu_4 = 0$, hence corresponding to the reaction schemes and reaction standard state free energies

$$\nu_1 C_1 + \nu_2 C_2 \rightleftharpoons \nu_3 C_3 \tag{3.74}$$

$$\Delta\mu^{\ominus} = \nu_3 \mu_3^{\ominus} - \nu_1 \mu_1^{\ominus} - \nu_2 \mu_2^{\ominus} \tag{3.75}$$

and

$$\nu_1 C_1 \rightleftharpoons \nu_3 C_3 \tag{3.76}$$

$$\Delta\mu^{\ominus} = \nu_3 \mu_3^{\ominus} - \nu_1 \mu_1^{\ominus} \tag{3.77}$$

respectively, we obtain (for the vapor–gas and solutes reactions)

$$e^{-\beta\Delta\mu^{\ominus}} = \frac{\alpha_3^{\nu_3}}{\alpha_1^{\nu_1}\alpha_2^{\nu_2}} \frac{(p_3/p^{\ominus})^{\nu_3}}{(p_1/p^{\ominus})^{\nu_1}(p_2/p^{\ominus})^{\nu_2}} = K_{eq} \tag{3.78}$$

$$e^{-\beta\Delta\mu^{\ominus}} = \frac{\alpha_3^{\nu_3}}{\alpha_1^{\nu_1}\alpha_2^{\nu_2}} \frac{(\rho_3/\rho^{\ominus})^{\nu_3}}{(\rho_1/\rho^{\ominus})^{\nu_1}(\rho_2/\rho^{\ominus})^{\nu_2}} = K_{eq} \tag{3.79}$$

for the $\nu_4 = 0$ case and

$$e^{-\beta\Delta\mu^{\ominus}} = \frac{\alpha_3^{\nu_3}}{\alpha_1^{\nu_1}} \frac{(p_3/p^{\ominus})^{\nu_3}}{(p_1/p^{\ominus})^{\nu_1}} = K_{eq} \tag{3.80}$$

$$e^{-\beta\Delta\mu^{\ominus}} = \frac{\alpha_3^{\nu_3}}{\alpha_1^{\nu_1}} \frac{(\rho_3/\rho^{\ominus})^{\nu_3}}{(\rho_1/\rho^{\ominus})^{\nu_1}} = K_{eq} \tag{3.81}$$

for the $\nu_2 = \nu_4 = 0$ case. Finally, considering again approximately unitary activity coefficients, the vapor–gas and solutes equilibrium constants reduce to

$$K_{eq} \cong \frac{(p_3/p^{\ominus})^{\nu_3}}{(p_1/p^{\ominus})^{\nu_1}(p_2/p^{\ominus})^{\nu_2}} \tag{3.82}$$

$$K_{eq} \cong \frac{(\rho_3/\rho^{\ominus})^{\nu_3}}{(\rho_1/\rho^{\ominus})^{\nu_1}(\rho_2/\rho^{\ominus})^{\nu_2}} \tag{3.83}$$

for the reaction scheme (3.74) and to

$$K_{eq} \cong \frac{(p_3/p^{\ominus})^{\nu_3}}{(p_1/p^{\ominus})^{\nu_1}} \tag{3.84}$$

$$K_{eq} \cong \frac{(\rho_3/\rho^{\ominus})^{\nu_3}}{(\rho_1/\rho^{\ominus})^{\nu_1}} \tag{3.85}$$

for the reaction scheme (3.76).

3.7 The Conformational Equilibrium

When considering for the jth solute the $A \rightleftharpoons B$ conformational transition, as defined by a semiclassical internal coordinate represented in the added jth solute molecule by the conformational coordinate ϱ_c with $\boldsymbol{\varrho}_{int} = \{\varrho_c, \boldsymbol{\varrho}'_{int}\}$, we can treat the two conformations as two different chemical species with thus at the thermodynamic equilibrium condition the equality of the corresponding chemical potentials, i.e., $\mu_{j_A} = \mu_{j_B}$. From Eq. 3.32 we have

$$\mu_{j_A} \cong -k_B T \ln\Big[\frac{\int_A e^{-\mathcal{H}_0(N_{j_A},N_{j_B})/(k_B T)}\, e^{-\Delta\mathcal{U}^0_{e,0}/(k_B T)}\, \sqrt{\det(\widetilde{m}_j)}\, d\boldsymbol{\xi} d\boldsymbol{\pi}_\xi d\varrho_c d\boldsymbol{\varrho}'_{int}/h^{d_j}}{\int e^{-\mathcal{H}_0(N_{j_A},N_{j_B})/(k_B T)}\, d\boldsymbol{\xi} d\boldsymbol{\pi}_\xi}\Big]$$
$$- k_B T \ln\Big[\big(2\pi k_B T\big)^{d_j/2}\Big] - k_B T \ln\Big[\frac{n_j\, q_{v,0,j}\, 8\pi^2}{(1+\gamma_j)}\Big] + k_B T \ln \rho_{j_A} \tag{3.86}$$

$$\mu_{j_B} \cong -k_B T \ln\Big[\frac{\int_B e^{-\mathcal{H}_0(N_{j_A},N_{j_B})/(k_B T)}\, e^{-\Delta\mathcal{U}^0_{e,0}/(k_B T)}\, \sqrt{\det(\widetilde{m}_j)}\, d\boldsymbol{\xi} d\boldsymbol{\pi}_\xi d\varrho_c d\boldsymbol{\varrho}'_{int}/h^{d_j}}{\int e^{-\mathcal{H}_0(N_{j_A},N_{j_B})/(k_B T)}\, d\boldsymbol{\xi} d\boldsymbol{\pi}_\xi}\Big]$$
$$- k_B T \ln\Big[\big(2\pi k_B T\big)^{d_j/2}\Big] - k_B T \ln\Big[\frac{n_j\, q_{v,0,j}\, 8\pi^2}{(1+\gamma_j)}\Big] + k_B T \ln \rho_{j_B} \tag{3.87}$$

where the A or B subscript of the integral sign means that integration over the ϱ_c semiclassical coordinate is confined within the A or B conformational range only, $N_{j_A} N_{j_B}$ are the molecular numbers of the A and B conformations before inserting the added molecule (conceived as corresponding to different chemical species and thus fixed in the integrals, with $N_{j_A} + N_{j_B} = N_j$) and we used $q_{v,0,j_A} \cong q_{v,0,j_B} \cong q_{v,0,j}$ (n_j, γ_j, d_j are clearly the same for all the jth species conformations). Therefore, from the equilibrium condition we must have

$$\mu_{j_B} - \mu_{j_A} \cong -k_B T \ln\Big[\frac{\int_B e^{-\mathcal{H}_0(N_{j_A},N_{j_B})/(k_B T)}\, e^{-\Delta\mathcal{U}^0_{e,0}/(k_B T)}\, \sqrt{\det(\widetilde{m}_j)}\, d\boldsymbol{\xi} d\boldsymbol{\pi}_\xi d\varrho_c d\boldsymbol{\varrho}'_{int}}{\int_A e^{-\mathcal{H}_0(N_{j_A},N_{j_B})/(k_B T)}\, e^{-\Delta\mathcal{U}^0_{e,0}/(k_B T)}\, \sqrt{\det(\widetilde{m}_j)}\, d\boldsymbol{\xi} d\boldsymbol{\pi}_\xi d\varrho_c d\boldsymbol{\varrho}'_{int}}\Big]$$
$$+ k_B T \ln\Big[\frac{\rho_{j_B}}{\rho_{j_A}}\Big] = 0 \tag{3.88}$$

and hence

$$\frac{\int_B e^{-\mathcal{H}_0(N^{eq}_{j_A},N^{eq}_{j_B})/(k_B T)}\, e^{-\Delta\mathcal{U}^0_{e,0}/(k_B T)}\, \sqrt{\det(\widetilde{m}_j)}\, d\boldsymbol{\xi} d\boldsymbol{\pi}_\xi d\varrho_c d\boldsymbol{\varrho}'_{int}}{\int_A e^{-\mathcal{H}_0(N^{eq}_{j_A},N^{eq}_{j_B})/(k_B T)}\, e^{-\Delta\mathcal{U}^0_{e,0}/(k_B T)}\, \sqrt{\det(\widetilde{m}_j)}\, d\boldsymbol{\xi} d\boldsymbol{\pi}_\xi d\varrho_c d\boldsymbol{\varrho}'_{int}}$$
$$= \frac{N^{eq}_{j_B}}{N^{eq}_{j_A}} \cong \frac{[B]_{eq}}{[A]_{eq}} = \frac{Q_B}{Q_A} \tag{3.89}$$

$$\frac{Q_B}{Q_A} \cong \frac{\int_B e^{-\mathcal{H}_0(N_j)/(k_B T)}\, e^{-\Delta\mathcal{U}^0_{e,0}/(k_B T)}\, \sqrt{\det(\widetilde{m}_j)}\, d\boldsymbol{\xi} d\boldsymbol{\pi}_\xi d\varrho_c d\boldsymbol{\varrho}'_{int}}{\int_A e^{-\mathcal{H}_0(N_j)/(k_B T)}\, e^{-\Delta\mathcal{U}^0_{e,0}/(k_B T)}\, \sqrt{\det(\widetilde{m}_j)}\, d\boldsymbol{\xi} d\boldsymbol{\pi}_\xi d\varrho_c d\boldsymbol{\varrho}'_{int}} \tag{3.90}$$

where $\rho_{j_B} = N_{j_B}^{eq}/V$ and $\rho_{j_A} = N_{j_A}^{eq}/V$ are the thermodynamic equilibrium molecular densities of the B and A conformations corresponding to the Helmholtz free energy minimum in the N_{j_A}, N_{j_B} space (see the first section of this chapter), $[B]_{eq} = \langle N_{j_B}/\mathcal{N}^0\rangle/V$ and $[A]_{eq} = \langle N_{j_A}/\mathcal{N}^0\rangle/V$ are the B and A equilibrium average molar densities (equilibrium concentrations) with $\mathcal{N}^0$ Avogadro's number and we used $\langle N_{j_A}\rangle \cong N_{j_A}^{eq}$, $\langle N_{j_B}\rangle \cong N_{j_B}^{eq}$ as following from the narrow Gaussian-like distribution of the N_{j_A} and N_{j_B} equilibrium fluctuations (see the first section of this chapter).

Furthermore, $\mathcal{Q}_B, \mathcal{Q}_A$ are the canonical partition functions of the system when allowing the N_j solute molecules to fluctuate between the A and B conformations and constraining the added molecule to be confined either in the B or in the A ϱ_c range: i.e., A and B are now conceived as the conformations of the same chemical species (N_{j_A}, N_{j_B} can change with $N_{j_A} + N_{j_B} = N_j$ fixed in the integrals of Eq. 3.90) and

$$P_B = \frac{\mathcal{Q}_B}{\mathcal{Q}_A + \mathcal{Q}_B} \tag{3.91}$$

$$P_A = \frac{\mathcal{Q}_A}{\mathcal{Q}_A + \mathcal{Q}_B} \tag{3.92}$$

provide the B and A equilibrium probabilities of any single jth solute molecule, with thus

$$\begin{aligned}\frac{P_B}{P_A} = \frac{\mathcal{Q}_B}{\mathcal{Q}_A} &\cong \frac{\int_B e^{-\mathcal{H}_0(N_j)/(k_BT)}\, e^{-\Delta\mathcal{U}_{e,0}^0/(k_BT)}\, \sqrt{\det(\widetilde{m}_j)}\, d\boldsymbol{\xi} d\boldsymbol{\pi}_\xi d\varrho_c d\boldsymbol{\varrho}'_{int}}{\int_A e^{-\mathcal{H}_0(N_j)/(k_BT)}\, e^{-\Delta\mathcal{U}_{e,0}^0/(k_BT)}\, \sqrt{\det(\widetilde{m}_j)}\, d\boldsymbol{\xi} d\boldsymbol{\pi}_\xi d\varrho_c d\boldsymbol{\varrho}'_{int}} \\ &\cong \frac{\int_B e^{-\mathcal{H}_0(N_{j_A}^{eq}, N_{j_B}^{eq})/(k_BT)}\, e^{-\Delta\mathcal{U}_{e,0}^0/(k_BT)}\, \sqrt{\det(\widetilde{m}_j)}\, d\boldsymbol{\xi} d\boldsymbol{\pi}_\xi d\varrho_c d\boldsymbol{\varrho}'_{int}}{\int_A e^{-\mathcal{H}_0(N_{j_A}^{eq}, N_{j_B}^{eq})/(k_BT)}\, e^{-\Delta\mathcal{U}_{e,0}^0/(k_BT)}\, \sqrt{\det(\widetilde{m}_j)}\, d\boldsymbol{\xi} d\boldsymbol{\pi}_\xi d\varrho_c d\boldsymbol{\varrho}'_{int}}\end{aligned} \tag{3.93}$$

Note that from Eqs. 3.81 and 3.89 we can write

$$-k_BT \ln \frac{\mathcal{Q}_B}{\mathcal{Q}_A} \cong \mu_{j_B}^{\ominus} - \mu_{j_A}^{\ominus} + k_BT \ln \frac{\alpha_{j_B}}{\alpha_{j_A}} \tag{3.94}$$

and hence when dealing with unitary activity coefficients for both the A and B conformations (i.e., $\alpha_{j_A} = \alpha_{j_B} = 1$) we have

$$-k_BT \ln \frac{\mathcal{Q}_B}{\mathcal{Q}_A} \cong \mu_{j_B}^{\ominus} - \mu_{j_A}^{\ominus} \tag{3.95}$$

$$\frac{[B]_{eq}}{[A]_{eq}} \cong K_{eq} \tag{3.96}$$

3.8 Examples

In this final section we show two examples, somehow representative of frequently addressed issues in contemporary soft matter computational chemistry: (i) the thermodynamics of reactive events in solution and (ii) the thermodynamics of conformational equilibria of complex (biomolecular) systems. A further example, dealing with time-dependent phenomena, will be reported at the end of the next chapter. Before entering into the details, we wish to preliminarily underline two important aspects. First of all, the reported applications of the equations previously outlined, although accurately described, are not supported by the technical–practical details for which we refer interested readers to duly referenced literature. Secondly, and most importantly, the spirit of the reported applications is the maintenance—as much as possible—of the physical coherence contained in the theoretical framework.

3.8.1 *Acetic Acid pKa*

Our goal is to calculate the equilibrium constant for the proton transfer reaction in aqueous solution from a Broensted acid molecule to a water molecule. For the calculation of this quantity, also known as K_a (or $pKa = -\log K_a$ using the more popular logarithmic notation) we have selected the simple deprotonation reaction of the Acetic acid (CH_3COOH) in aqueous solution, with the pKa defined as

$$pKa \cong \frac{\mu^{\ominus}_{H^+} + \mu^{\ominus}_{Ac^-} - \mu^{\ominus}_{AcH}}{2.303\; k_B T} \tag{3.97}$$

according to the reaction scheme $AcH \rightleftharpoons Ac^- + H^+$ (see the equilibrium constant section), where the key quantities to model are the standard chemical potentials of all the involved species in aqueous solution: the Acetic acid (AcH), the Acetate anion (Ac^-) and the proton H^+. The standard chemical potential of the aqueous proton can be estimated by the equation

$$\mu^{\ominus}_{H^+} = \Delta G_{H^+} + k_B T \ln\left(k_B T \rho^{\ominus}/p^{\ominus}\right) + \mu^{\ominus}_{H^+,gas} \tag{3.98}$$

where the terms on the right side of the equation correspond (from left to right) to the solvation free energy of the proton in water equal to -1082 kJ/mol [3] at 298 K, the correction term for the change of the standard state from the gas ($p^{\ominus} = 0.1$ MPa) to solution ($\rho^{\ominus} = 1.0$ M) phase equal to 7.9 kJ/mole at 298 K and, finally, the gaseous proton standard chemical potential. By using Eqs. 3.42–3.45 we can obtain a simple expression of $\mu^{\ominus}_{Ac^-,gas} - \mu^{\ominus}_{AcH,gas}$ (the $AcH \rightarrow Ac^-$ gas-phase standard state chemical potential change) when realizing that both chemical species can be treated as rigid or semirigid molecules with thus only vibrational internal coordinates, treated by the molecular vibrational partition function, and the molecular roto-translational coordinates as the solely explicit semiclassical degrees of freedom. Note that for the dihedral angle in the Acetyl group (a semiclassical coordinate at room temperature) we must consider a single energy minimum basin, as the other two equivalent minima correspond to permutations of identical particles (i.e., the methyl hydrogens) hence providing physically indistinguishable semiclassical configurations. Therefore, also such a dihedral angle can be considered as a vibrational harmonic-like coordinate to be treated by the vibrational partition function, with thus no left explicit semiclassical internal coordinates. Both molecules have then an identical number of semiclassical degrees of freedom, i.e., the roto-translational ones, with a fixed energy change $\Delta\mathcal{U}$ in Eq. 3.42 and fully constant mass tensors involving the 3×3 diagonal center of mass block, with the non-zero diagonal elements given by the molecular mass, and the 3×3 inertia tensor corresponding to the rotational angular velocity (see Sect. 3.3). Employing such considerations in Eqs. 3.42–3.45 to obtain $\mu^{\ominus}_{Ac^-,gas} - \mu^{\ominus}_{AcH,gas}$ we can express the gas-phase standard reaction free energy $\Delta\mu^{\ominus}_{gas} = \mu^{\ominus}_{Ac^-,gas} - \mu^{\ominus}_{AcH,gas} + \mu^{\ominus}_{H^+,gas}$ (gas-phase basicity, 1427 kJ/mol at 298 K from experimental data [4]) via

$$\Delta\mu^{\ominus}_{gas} \cong \Delta\mathcal{U}_{gas} + \Delta A^{gas}_{v} + \mu^{\ominus}_{H^+,gas} \tag{3.99}$$

$$\begin{aligned}\Delta A^{gas}_{v} &= -k_B T \ln\left[\frac{q_{v,0,Ac^-}}{q_{v,0,AcH}}\right] - k_B T \ln\left[\frac{n_{Ac^-}(1+\gamma_{AcH})}{n_{AcH}(1+\gamma_{Ac^-})}\right] \\ &- \frac{k_B T}{2}\ln\left[\frac{\det(\widetilde{m}_{Ac^-})}{\det(\widetilde{m}_{AcH})}\right]\end{aligned} \tag{3.100}$$

where $\Delta\mathcal{U}_{gas}$ is the $AcH \rightarrow Ac^-$ gas-phase electronic ground eigenstate energy change, as obtained by using the ground eigenenergies of AcH and Ac^- at the corresponding (vacuum) optimized geometries (1426 kJ/mol from quantum chemical calculations), and ΔA^{gas}_{v} is the vibrational–rotational contribution (i.e., including all the terms beyond the first one in Eq. 3.42) to the gas-phase free energy change.

From the definition of the chemical potential difference (by using the Helmholtz free energy change, see Eq. 3.41) and employing again Eqs. 3.42–3.45 (within the high dilution limit) we can write for the $AcH \rightarrow Ac^-$ transition in aqueous solution

$$\mu^{\ominus}_{Ac^-} - \mu^{\ominus}_{AcH} \cong \Delta A_{el} + \Delta A^{aq}_{v} \tag{3.101}$$

where ΔA^{aq}_{v} is the vibrational–rotational contribution to the aqueous solution $AcH \rightarrow Ac^-$ free energy change, as provided by the same expression of Eq. 3.100 with now the solution molecular vibrational partition functions. Moreover, ΔA_{el} is the deprotonation electronic free energy term given by

$$\begin{aligned}\Delta A_{el} &\cong -k_B T \ln\langle e^{-\Delta\mathcal{U}/(k_B T)}\rangle_{AcH} + \Delta A^{ion}_{dep} \\ &= k_B T \ln\langle e^{\Delta\mathcal{U}/(k_B T)}\rangle_{Ac^-} - \Delta A^{ion}_{prot}\end{aligned} \tag{3.102}$$

with $\Delta\mathcal{U}$ the aqueous solution electronic ground eigenstate energy difference for the $AcH \rightarrow Ac^-$ transition (the transition energy now fluctuating according to the solvent perturbation) as obtained by using the perturbed ground eigenenergies of AcH and Ac^- at the corresponding (vacuum) optimized geometries [5], the subscript of the angle brackets clearly indicates the chemical state statistical ensemble used for averaging over the transition energy fluctuations, ΔA^{ion}_{dep} is the free energy change due to the ionic environment relaxation following the $AcH \rightarrow Ac^-$ transition and, similarly, ΔA^{ion}_{prot} is the free energy change due to the ionic environment relaxation following the backward transition $Ac^- \rightarrow AcH$. It is reasonable to consider these two relaxation free energies rather similar, then allowing us to assume $\Delta A^{ion}_{dep} \approx \Delta A^{ion}_{prot}$ and thus from Eq. 3.102 it follows

$$\Delta A_{el} \cong \frac{k_B T}{2} \ln \frac{\langle e^{\Delta\mathcal{U}/(k_B T)} \rangle_{Ac^-}}{\langle e^{-\Delta\mathcal{U}/(k_B T)} \rangle_{AcH}} \tag{3.103}$$

By using isochoric (N, V, T) Molecular Dynamics (MD) simulations of AcH and Ac^-, each embedded into a large water box mimicking the experimental aqueous solution condition at high solute dilution [6], see Fig. 3.1, and employing quantum-classical calculations to obtain the transition energy at each MD frame [7], we obtained the proper two statistical ensembles and related expectation values to be used in the last equation, providing $\Delta A_{el} = 1100$ kJ/mol.

Finally, considering the little effects of the solvent perturbation on the vibrational frequencies, we can safely assume $\Delta A^{aq}_v \cong \Delta A^{gas}_v$ and thus from Eqs. 3.97, 3.98 and 3.101 we have

$$pKa \cong \frac{\Delta A_{el} + \Delta G_{H^+} + \Delta\mu^{\ominus}_{gas} + k_B T \ln\left(k_B T \rho^{\ominus}/p^{\ominus}\right) - \Delta\mathcal{U}_{gas}}{2.303\, k_B T} \tag{3.104}$$

providing, when inserting all the values obtained from experimental and computational data, $pKa = 4.7$ in excellent agreement with the experimentally measured 4.8 value.

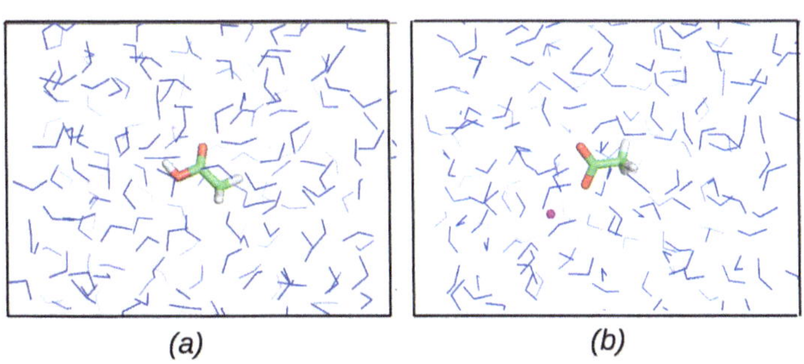

Fig. 3.1 Schematic view of the simulation box for AcH (panel a), and Na^+/Ac^- (panel b)

3.8.2 Di-Alanine Conformational Equilibrium

In this second example we address the thermodynamics of a solute conformational equilibrium as defined by a solute internal conformational coordinate, ϱ_c with $\boldsymbol{\varrho}_{int} = \{\varrho_c, \boldsymbol{\varrho}'_{into}\}$ (see Sect. 3.7). For a diluted solute from Eqs. 3.93 and 3.96 (omitting the j index as we now consider a single solute species) we can easily obtain the standard chemical potential difference of the A and B conformations via

$$-k_B T \ln \frac{P_B}{P_A} \cong \mu_B^{\ominus} - \mu_A^{\ominus} \tag{3.105}$$

where P_A and P_B (see Sect. 3.7) are the equilibrium probabilities of the two conformations (i.e., the equilibrium probabilities for a single solute molecule identical to the equilibrium average fractions of the A and B conformations within the solute molecules). This is a very important and widely investigated issue in Computational Chemistry and Biophysics [8, 9] because of the tight connection between conformational equilibria and (bio)chemical properties of peptides, proteins and, more in general, biomolecular systems. The starting necessary point of (basically) all the investigations devoted to this purpose is an exhaustive sampling of the conformational repertoire of the molecular system of interest, carried out with semiclassical [10] time-dependent (MD simulations) or stochastic (Monte Carlo) methods [11]. The second, and generally more complicated issue, is the identification of the internal conformational coordinates, "essential" for tackling the conformational equilibrium. We do not address this important point in this book referring the reader to dedicated literature [12]. In the example we report here the conformational equilibrium of a simple solvated dipeptide is investigated, by employing the straightforward Ramachandran ϕ and ψ angles (see Fig. 3.2) to define the proper conformational space. The selected system, shown in Fig. 3.3, is the "capped" di-Alanine (hereafter di-ala) in water which was simulated by MD using a rigid box [6] with one di-ala molecule embedded in a large amount of explicit water molecules at 338 K. Such a relatively high temperature is preferred to the typical room temperature ($\approx$ 300 K) because it conceivably allows us to span a wider conformational repertoire with a faster statistical sampling, making the conformational analysis more interesting for our didactic purposes. The simulation performed provided a 100 ns MD trajectory, reasonably extended for the simple system considered to ensure a proper equilibrium statistics under the usual assumption of the ergodic behavior, i.e., each MD trajectory within the large time limit provides a time-distribution of any observable converging to the corresponding equilibrium ensemble distribution. Therefore, once evaluated at each MD frame the instantaneous ϕ and ψ values of di-ala, we constructed their distribution over the Ramachandran plane as reported in Fig. 3.4. From this figure it is evident that we obtained eight conformational basins (indicated in the figure by the red boxes) corresponding to the thermodynamically stable conformational regions. Subdividing the Ramachandran plane of Fig. 3.4 into a large number of grid cells,

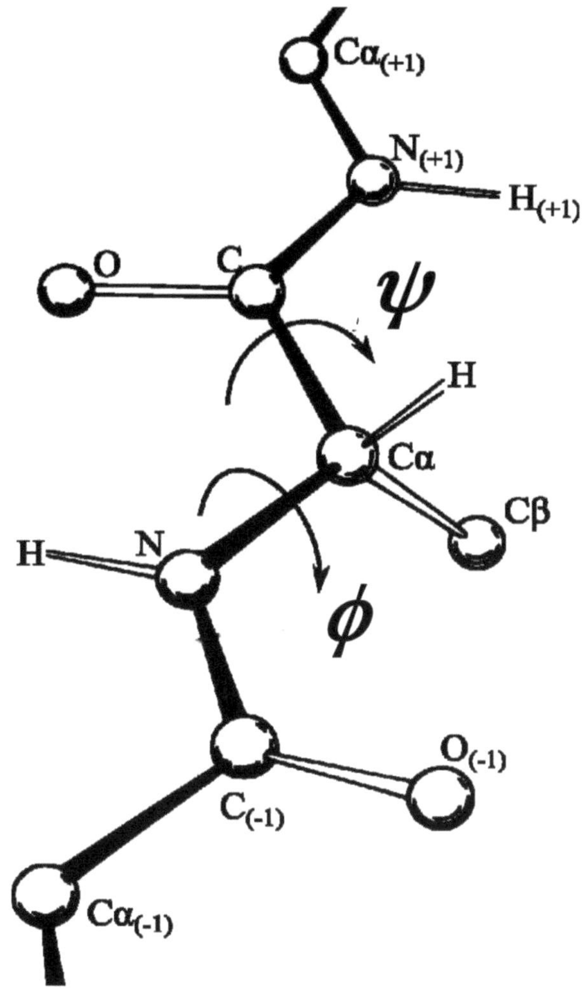

Fig. 3.2 Ramachandran coordinates

each defining a conformation within the ϕ, ψ space, we obtained the corresponding probabilities via

$$P_{\phi_i,\psi_i} = \frac{N_i}{N_{frames}} \tag{3.106}$$

where ϕ_i, ψ_i is the Ramachandran configuration at the center of the ith grid cell, N_i is the number of MD frames found within the ith grid cell and N_{frames} is the total number of MD frames. Such grid cell probabilities can be then used to reconstruct the standard chemical potential (discrete) surface of the grid cell conformational states by

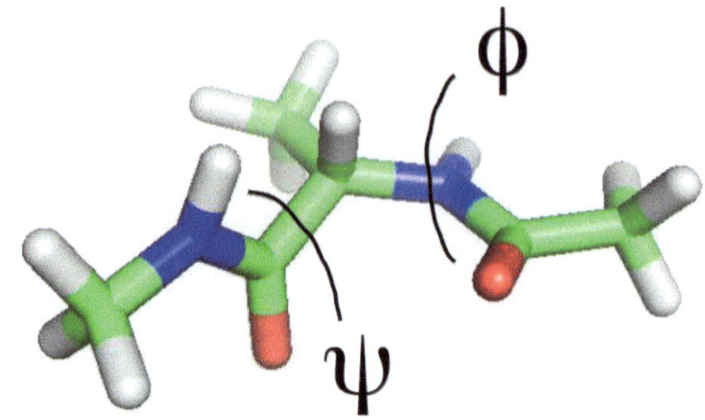

Fig. 3.3 Schematic view of capped di-Alanine showing the Ramachandran coordinates

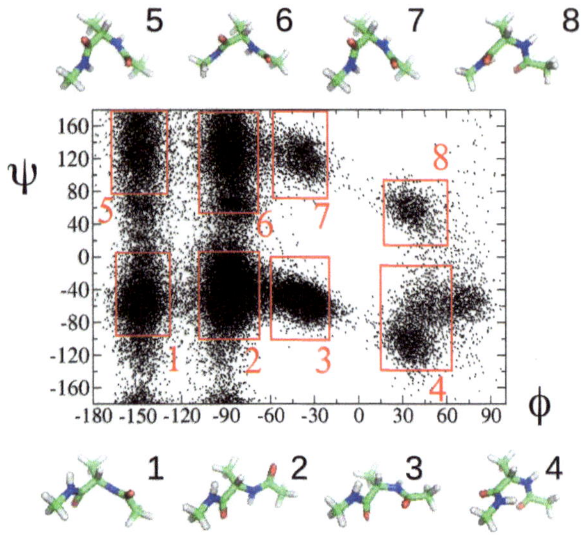

Fig. 3.4 The ϕ, ψ configurations collected during the MD simulation. The red boxes indicate the high probability regions involving a high density of configurations (for each region a representative structure is shown)

$$\Delta\mu^{\ominus}_{\phi_i,\psi_i} = \mu^{\ominus}_{\phi_i,\psi_i} - \mu^{\ominus}_{\phi_r,\psi_r} \cong -k_B T \ln \frac{P_{\phi_i,\psi_i}}{P_{\phi_r,\psi_r}} \tag{3.107}$$

utilizing the grid cell with the highest probability (centered in ϕ_r, ψ_r) as the reference conformational state and resulting in the Ramachandran free energy landscape, reported in Fig. 3.5, clearly showing the eight conformational basins as well separated free energy minima. It is worth noting that the conformational free energy surface obtained, beyond providing quantitative information on the relevant conformational regions and their interconnecting barriers, provides the basis for further evaluating the conformational thermodynamics (e.g., the Ramachandran grid cell enthalpy and entropy) and may furnish the necessary information for reconstructing the kinetics of conformational transitions.

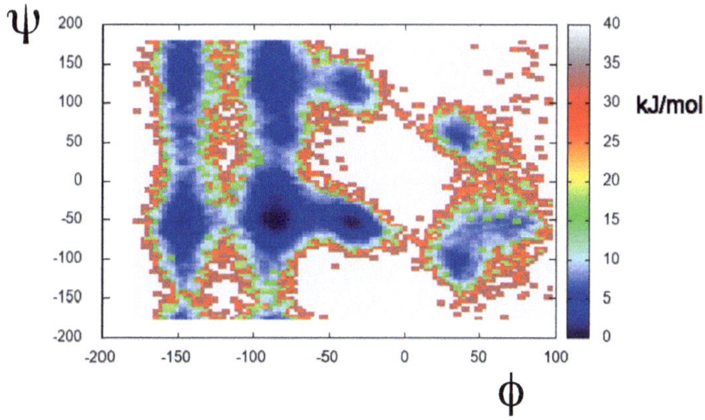

Fig. 3.5 The free energy landscape in the Ramachandran plane as obtained by Eq. 3.107 and MD data

References

1. Gallavotti G (1983) The elements of mechanics. Springer, New York
2. Amadei A, Chillemi G, Ceruso MA, Grottesi A, Di Nola A (2000) Molecular simulations with constrained roto-translational motions: theoretical basis and statistical mechanical consistency. J Chem Phys 112:9–23
3. Hunenberger P, Reif M (2011) Single-ion solvation: experimental and theoretical approaches to elusive thermodynamic quantities. Royal Society of Chemistry Publishing
4. Taft RW, Topsom RD (1987) The nature and analysis of substituent electronic effects. In: Taft RW (ed) Progress in physical organic chemistry. Wiley, p 1–83
5. Amadei A, D'Alessandro M, D'Abramo M, Aschi M (2009) Theoretical characterization of electronic states in interacting chemical systems. J Chem Phys 130:084109
6. The size of the box is preventively adjusted in order to reproduce the same (average) pressure observed by simulating a pure water box at the experimental density
7. Zanetti-Polzi L, Daidone I, Amadei A (2020) Fully atomistic multiscale approach for pK_a prediction. J Phys Chem B 124:4712
8. Dror RO, Dirks RM, Grossman J, Xu H, Shaw DE (2012) Biomolecular simulation: a computational microscope for molecular biology. Annu Rev Biophys 41:429
9. Adcock SA, McCammon JA (2006) Molecular dynamics: survey of methods for simulating the activity of proteins. Chem. Rev. 106:1589
10. With the term semiclassical simulations we indicate all the methods, nowadays available and making use of either the empirical force-fields or based on Quantum-Chemical calculations, treating the nuclei as 'classical' particles subject to the Newton equations of motion
11. Robert CP, Casella G (2004) Monte Carlo statistical methods. Springer, New York
12. Daidone I, Amadei A (2012) Essential dynamics: foundation and applications. WIREs Comput Mol Sci 2:762

Chapter 4
Statistical Mechanics: Application to Chemical Kinetics

Abstract In this final chapter, statistical mechanics is used to develop general theoretical models for the time dependence of chemical transformations within the chemical kinetics framework. After introducing the Landau free energy, the expression for the unimolecular rate constant is derived and its relationship with the popular Eyring equation is illustrated. Subsequently, a general model for the treatment of bimolecular reactions is outlined in detail. Finally, an example is reported dealing with the kinetics of a paradigmatic reaction in organic chemistry: the bimolecular nucleophile substitution reaction in solution.

4.1 A General Model for Unimolecular Chemical Reactions

Let us consider a simple reaction from the reactant (R) chemical state to the product (P) chemical state, without involving any distinguishable intermediate chemical state, as occurring within a reactive center (RC), either a molecule or a complex, interacting with its environment. Note that we assume that the reaction rate constants be invariant during the whole reaction process and for the R and P chemical states a full inner equilibrium, i.e., the corresponding mean residence time is large enough to allow for internal relaxation. We can also reasonably assume that the reaction events must pass through a specific and unique (high free energy) transition state (TS) with a mean residence time too short for any internal relaxation, hence providing the reaction scheme

$$R \xrightarrow{\mathcal{K}_R} TS \xrightarrow{k} P \tag{4.1}$$

and thus

$$[\dot{R}] = -\mathcal{K}_R[R] \tag{4.2}$$

$$[\dot{TS}] = \mathcal{K}_R[R] - k[TS] \tag{4.3}$$

$$[\dot{P}] = k[TS] \tag{4.4}$$

A. Amadei and M. Aschi, *Statistical Mechanics for Chemical Thermodynamics and Kinetics*, https://doi.org/10.1007/978-3-031-77929-9_4

with $[X]$ the concentration of the X species (i.e., the molar density). Considering that we necessarily have $k \gg \mathcal{K}_R$ we can assume the steady state approximation for TS (i.e., $[\dot{TS}] \cong 0$) providing

$$[TS] \cong \frac{\mathcal{K}_R}{k}[R] \tag{4.5}$$

$$[\dot{P}] \cong k\frac{\mathcal{K}_R}{k}[R] = \mathcal{K}_R[R] \tag{4.6}$$

It is reasonable to assume that within the TS equilibrium ensemble the outward flux rate constants be identical (due to the TS tiny mean residence time) and given by $k_{TS} \cong k/2$ (at equilibrium only one half of the TS population is moving toward P), and thus realizing that the $R \rightleftharpoons TS$ equilibrium is equivalent to a conformational one, from Eq. 3.89 of Chap. 3, we have

$$\frac{\mathcal{K}_R}{k_{TS}} = \frac{[TS]_{eq}}{[R]_{eq}} = \frac{Q_{TS}}{Q_R} \tag{4.7}$$

with $[TS]_{eq}$, $[R]_{eq}$ the equilibrium TS and R state concentrations and Q_{TS}, Q_R the canonical partition functions of the TS and R state, respectively, as obtained for the whole RC-environment system when considering a selected (single) reference RC confined either within the TS or within the R chemical state (for reactive centers at high dilution we can obtain such partition functions by using a single RC embedded in a vast amount of solvent molecules). Therefore, we can write

$$\begin{aligned}[\dot{P}] &\cong -[\dot{R}] = \mathcal{K}_R[R] = \frac{\mathcal{K}_R}{k_{TS}} k_{TS}[R] \\ &= \frac{Q_{TS}}{Q_R} k_{TS}[R] \cong \frac{Q_{TS}}{Q_R}\frac{k}{2}[R] \end{aligned} \tag{4.8}$$

$$\mathcal{K}_R \cong \frac{Q_{TS}}{Q_R}\frac{k}{2} \tag{4.9}$$

4.1.1 The Reaction Coordinate and the Landau Free Energy

Let us assume that we can identify for the reference RC a generalized semiclassical coordinate ξ (i.e., the reaction coordinate) univocally defining via its different intervals the R, TS and P chemical states and thus properly describing the reaction process. Note that the reaction coordinate is defined to have the R state corresponding to the lowest ξ values range. Defining with $[\xi_{TS} - \delta/2, \xi_{TS} + \delta/2]$ the TS tiny reaction coordinate range and with ξ_m the minimum ξ value of the R chemical state, we can express the partition functions Q_R and Q_{TS} via the Landau free energy [1] $\mathcal{A}$ (see Fig. 4.1) defined by

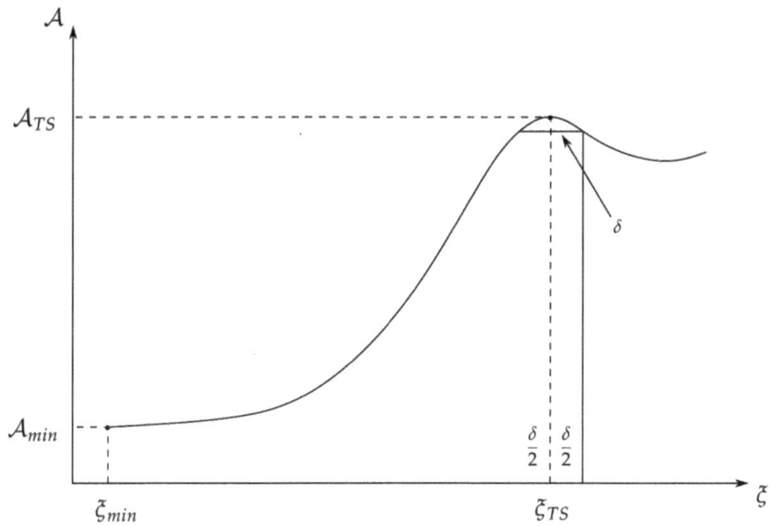

Fig. 4.1 Pictorial representation of a generic Landau free energy profile along the reaction coordinate ξ

$$\mathcal{A}(\xi) = -k_B T \ln \mathcal{Q}(\xi) \tag{4.10}$$

with $\mathcal{Q}(\xi)$ the partition function density providing

$$Q_R = \int_{\xi_m}^{\xi_{TS}-\delta/2} \mathcal{Q}(\xi)d\xi \cong \int_{\xi_m}^{\xi_{TS}} \mathcal{Q}(\xi)d\xi = \int_{\xi_m}^{\xi_{TS}} e^{-\beta\mathcal{A}(\xi)}d\xi \tag{4.11}$$

$$Q_{TS} = \int_{\xi_{TS}-\delta/2}^{\xi_{TS}+\delta/2} \mathcal{Q}(\xi)d\xi \cong \mathcal{Q}(\xi_{TS})\delta = e^{-\beta\mathcal{A}(\xi_{TS})}\delta \tag{4.12}$$

where we used $\mathcal{A}(\xi) \cong \mathcal{A}(\xi_{TS})$ within the TS range. In practice, from the general expression of the canonical partition function (see Eqs. 3.22 and 3.23 of Chap. 3)

$$Q \cong \Gamma Q_{v,0} \int e^{-\beta\mathcal{H}_0(\xi,\zeta,\pi_\xi,\pi_\zeta)} \frac{d\zeta d\pi_\zeta d\xi d\pi_\xi}{h^D} \tag{4.13}$$

we can define the partition function density via

$$\begin{aligned}\mathcal{Q}(\xi) \cong\ & \Gamma Q_{v,0} \int e^{-\beta\mathcal{H}_0(\xi',\zeta,\pi_\xi,\pi_\zeta)} \delta(\xi' - \xi) \frac{d\zeta d\pi_\zeta d\xi' d\pi_\xi}{h^D} \\ =\ & \Gamma Q_{v,0} \int e^{-\beta\mathcal{H}_0(\xi,\zeta,\pi_\xi,\pi_\zeta)} \frac{d\zeta d\pi_\zeta d\pi_\xi}{h^D}\end{aligned} \tag{4.14}$$

where $\delta(\xi' - \xi)$ is the Dirac function, $1/\beta = k_B T$, $Q_{v,0}$ and $\mathcal{H}_0$ are the vibrational partition function and semiclassical Hamiltonian for the electronic ground state, Γ is a constant including the quantum correction for the permutations of identical particles, D is the total number of semiclassical degrees of freedom and $\boldsymbol{\zeta}, \boldsymbol{\pi}_\zeta$ are the semiclassical coordinates and conjugate momenta complementary to the reaction coordinate ξ and its conjugate momentum π_ξ (note that for any semiclassical coordinates and conjugate momenta transformation the Jacobean is always unitary). When considering chemical reactions as occurring in a given electronic excited eigenstate of the system (i.e., the reactions cannot occur in the electronic ground eigenstate), in Eq. 4.14 we clearly have to use the substitutions $Q_{v,0} \to Q_{v,j_e}$ and $\mathcal{H}_0 \to \mathcal{H}_{j_e}$ with $j_e > 0$.

From Eqs. 4.11 and 4.12 the equilibrium probability densities for the R and TS ensembles (ρ_R and ρ_{TS}) can be obtained

$$\rho_R(\xi) = \frac{\mathcal{Q}(\xi)}{\mathcal{Q}_R} = \frac{e^{-\beta\mathcal{A}(\xi)}}{\mathcal{Q}_R} \tag{4.15}$$

$$\xi_m \leq \xi \leq \xi_{TS} - \frac{\delta}{2} \tag{4.16}$$

$$\rho_{TS}(\xi) = \frac{\mathcal{Q}(\xi)}{\mathcal{Q}_{TS}} = \frac{e^{-\beta\mathcal{A}(\xi)}}{\mathcal{Q}_{TS}} \cong \frac{1}{\delta} \tag{4.17}$$

$$\xi_{TS} - \frac{\delta}{2} \leq \xi \leq \xi_{TS} + \frac{\delta}{2} \tag{4.18}$$

Furthermore, when we can assume that within the R range $\mathcal{A}(\xi)$ be well approximated via a quadratic function relevantly extended around its minimum, the reactant state Landau free energy is given by

$$\mathcal{A}_R(\xi) = \mathcal{A}(\xi_R) + \frac{k_B T}{2\sigma_R^2}\left(\xi - \xi_R\right)^2 \tag{4.19}$$

$$\xi_R \cong \langle\xi\rangle_R \tag{4.20}$$

$$\sigma_R^2 \cong \langle\left(\xi - \langle\xi\rangle_R\right)^2\rangle_R \tag{4.21}$$

with ξ_R the R Landau free energy minimum position (virtually identical to the R ensemble average ξ value $\langle\xi\rangle_R$) and σ_R^2 the ξ variance as provided by the quadratic function (virtually identical to the R ensemble variance $\langle\left(\xi - \langle\xi\rangle_R\right)^2\rangle_R$). Therefore, from Eqs. 4.11 and 4.19, assuming $(\xi_{TS} - \xi_m)/\sigma_R \geq 6$, we can write for such a Gaussian approximation

$$\mathcal{Q}_R = \int_{\xi_m}^{\xi_{TS}-\delta/2} e^{-\beta\mathcal{A}(\xi)} d\xi \cong \int_{-\infty}^{\infty} e^{-\beta\mathcal{A}_R(\xi)} d\xi$$
$$= e^{-\beta\mathcal{A}(\xi_R)}\sqrt{2\pi\sigma_R^2}. \tag{4.22}$$

4.1.2 The Rate Constant

From Eqs. 4.9, 4.11 and 4.12 we obtain

$$\mathcal{K}_R \cong \frac{e^{-\beta\Delta\mathcal{A}_R^\dagger}\delta}{\int_{\xi_m}^{\xi_{TS}} e^{-\beta\Delta\mathcal{A}(\xi)}d\xi}\frac{k}{2} \tag{4.23}$$

$$\Delta\mathcal{A}(\xi) = \mathcal{A}(\xi) - \mathcal{A}(\xi_R) \tag{4.24}$$

$$\Delta\mathcal{A}_R^\dagger = \mathcal{A}(\xi_{TS}) - \mathcal{A}(\xi_R) \tag{4.25}$$

providing a general expression for the reaction rate constant which, when the Gaussian approximation of Eq. 4.22 holds, can be simplified to

$$\mathcal{K}_R \cong \frac{e^{-\beta\Delta\mathcal{A}_R^\dagger}\delta}{\sqrt{2\pi\sigma_R^2}}\frac{k}{2} \tag{4.26}$$

Realizing that the TS, corresponding to a tiny ξ interval, is characterized by an approximately homogeneous density within the whole δ range (i.e., $\rho = [TS]/\delta$) and for each reaction event by a virtually constant crossing velocity, we can express the outward flux $k[TS]$ (forming the P population) via $\rho\langle v\rangle = [TS]\langle v\rangle/\delta$ with $\langle v\rangle$ the mean traversing velocity given by the equilibrium mean value of $v = |\dot{\xi}|$, readily providing

$$k \cong \frac{\langle v\rangle}{\delta} \tag{4.27}$$

and hence from Eq. 4.23

$$\mathcal{K}_R \cong \frac{e^{-\beta\Delta\mathcal{A}_R^\dagger}}{\int_{\xi_m}^{\xi_{TS}} e^{-\beta\Delta\mathcal{A}(\xi)}d\xi}\frac{\langle v\rangle}{2} \tag{4.28}$$

Note that due to quantum dynamical effects it is possible that only a fraction of the reactive population exiting the TS be transformed into the product chemical state (as often occurring in charge transfer reactions), thus possibly requiring in Eqs. 4.27 and 4.28 an extra factor corresponding to such a fraction (i.e., the transmission coefficient). In this book we do not specifically address such reactions and then, for sake of simplicity, we always assume a unitary transmission coefficient.

When considering the usual Gaussian-like probability for the equilibrium distribution of $\dot{\xi}$, it follows $\langle v\rangle = \langle|\dot{\xi}|\rangle \cong \sigma_{\dot{\xi}}/\sqrt{\pi/2}$ with $\sigma_{\dot{\xi}}^2$ the equilibrium ensemble variance of $\dot{\xi}$, thus providing

$$\mathcal{K}_R \cong \frac{e^{-\beta\Delta\mathcal{A}_R^\dagger}}{\int_{\xi_m}^{\xi_{TS}} e^{-\beta\Delta\mathcal{A}(\xi)}d\xi}\frac{\sigma_{\dot{\xi}}}{2\sqrt{\pi/2}} \tag{4.29}$$

or from Eq. 4.26 (i.e., assuming the Gaussian approximation of Eq. 4.22)

$$\mathcal{K}_R \cong \frac{e^{-\beta\Delta\mathcal{A}_R^\dagger}}{\sqrt{2\pi\sigma_R^2}} \frac{\sigma_{\dot{\xi}}}{2\sqrt{\pi/2}} \tag{4.30}$$

where for the typical ξ reaction coordinate corresponding to the negative eigenvalue eigenvector of the RC mass weighted Hessian at the electronic energy saddle point identifying the transition state, we can use the velocity variance of the eigenvector mass weighted coordinate, i.e., $\sigma_{\dot{\xi}} = \sqrt{k_B T}$ (note that it is usually assumed no significant change of the TS structure and mass weighted Hessian eigenvectors due to the RC environment perturbation). Moreover, when within the R range such a ξ reaction coordinate also corresponds to an uncoupled normal mode of an electronic energy harmonic well (clearly fulfilling Eq. 4.30), we have $\sigma_{\dot{\xi}}/\sigma_R = \omega_\xi = 2\pi\nu_\xi$ where ν_ξ is the mode frequency. Therefore, when considering any reaction coordinate fulfilling the Gaussian approximations (i.e., Eq. 4.30), we can define the corresponding *frequency* via the same relation and thus obtain

$$\mathcal{K}_R \cong \frac{e^{-\beta\Delta\mathcal{A}_R^\dagger}}{\sqrt{2\pi\sigma_R^2}} \frac{\sigma_{\dot{\xi}}}{2\sqrt{\pi/2}} = e^{-\beta\Delta\mathcal{A}_R^\dagger} \frac{\sigma_{\dot{\xi}}/\sigma_R}{2\pi} = e^{-\beta\Delta\mathcal{A}_R^\dagger}\,\nu_\xi \tag{4.31}$$

Finally, assuming $h\nu_\xi \approx k_B T$ with h Planck's constant (i.e., the reaction coordinate within the R range can be considered a stiff semiclassical *mode* coordinate) and realizing that $\Delta\mathcal{A}_R^\dagger \cong \Delta A_R^\dagger = A_{\xi_{TS}} - A_{\xi_R}$ we have

$$\mathcal{K}_R \approx e^{-\beta\Delta A_R^\dagger}\, k_B T/h \tag{4.32}$$

where

$$A_{\xi_{TS}} \cong \mathcal{A}(\xi_{TS}) - k_B T \ln\delta \tag{4.33}$$

$$A_{\xi_R} \cong \mathcal{A}(\xi_R) - k_B T \ln\delta \tag{4.34}$$

are the system free energies when constraining the reference RC within either the TS or within the reactant state Landau free energy minimum (i.e., the RC is confined within the δ range around the ξ_R position). Equation 4.32 is typically referred to as the Eyring equation providing a special case of the more general Eqs. 4.28–4.30.

All the approximations and derivations described in this section can be used to obtain also the $\mathcal{K}_P$ rate constant

$$\mathcal{K}_P \cong \frac{e^{-\beta\Delta\mathcal{A}_P^\dagger}}{\int_{\xi_{TS}}^{\xi_M} e^{-\beta\Delta\mathcal{A}(\xi)}d\xi} \frac{\langle v\rangle}{2} \tag{4.35}$$

$$\Delta\mathcal{A}(\xi) = \mathcal{A}(\xi) - \mathcal{A}(\xi_P) \tag{4.36}$$

$$\Delta\mathcal{A}_P^\dagger = \mathcal{A}(\xi_{TS}) - \mathcal{A}(\xi_P) \tag{4.37}$$

for the $P \xrightarrow{\mathcal{K}_P} TS \xrightarrow{k} R$ back reaction (ξ_M, ξ_P being the P upper bound and Landau free energy minimum position), thus providing the complete kinetics of the $R \rightleftharpoons P$ interconversion

$$[\dot{R}] \cong -\mathcal{K}_R[R] + \mathcal{K}_P[P] \tag{4.38}$$

$$[\dot{P}] \cong \mathcal{K}_R[R] - \mathcal{K}_P[P] \tag{4.39}$$

4.2 A Simple Model for Bimolecular Chemical Reactions

In this section we describe in detail a model for bimolecular reactions, firstly introduced by Hammes [2], describing a chemical complex formation.

Let us consider a simple molecular complex formation described by the following bimolecular reaction scheme

$$A + B \xrightarrow{k_A} C \tag{4.40}$$

where the three chemical species involved (A, B and C) are embedded into a fluid-state system with the reaction rate constant invariant during the whole reaction process. For a given single A molecule, we can conceive the corresponding C formation events as occurring any time a B molecule enters the r_0 radius sphere centered in the A molecule center of mass (r_0 is the distance of the A and B centers of mass). Therefore, the B flux experienced by a generic single A molecule, assuming a pure diffusive behavior of the B molecules for $r \geq r_0$ and using the A molecule reference frame (reference molecular frame) instead of the fixed laboratory one, can be expressed by the first Fick's law

$$\boldsymbol{J} = -D\nabla\rho_r \tag{4.41}$$

where $\boldsymbol{J}(t, \boldsymbol{r})$ is the flux density vector for the B molecules toward a single A molecule and $\rho_r(t, \boldsymbol{r})$ is the B molecular density, both a function of time t and of the Cartesian spatial position

$$\boldsymbol{r} = \begin{bmatrix} x \\ y \\ z \end{bmatrix} \tag{4.42}$$

as expressed within the reference molecular frame. Furthermore, D is the B molecules diffusion coefficient as obtained within the reference molecular frame and clearly (omitting in the notation the fixed variables)

$$\nabla\rho_r = \begin{bmatrix} \frac{\partial\rho_r}{\partial x} \\ \frac{\partial\rho_r}{\partial y} \\ \frac{\partial\rho_r}{\partial z} \end{bmatrix} \tag{4.43}$$

Note that the origin of the reference molecular frame is the A molecule center of mass and the B molecular density is obtained by averaging over all the A molecules the corresponding B molecules centers of mass distributions, and thus ρ_r is the mean B molecular density around a single A molecule as expressed within the reference molecular frame.

The B flux toward the A molecule at each radial distance $|\boldsymbol{r}| = r$ (i.e., across the r-radius spherical surface) is then

$$\begin{aligned}\Phi(t, r) &= \int \boldsymbol{J}(t, \boldsymbol{r}) \cdot \frac{\boldsymbol{r}}{r}\, d\sigma = -D \int \nabla\rho_r \cdot \frac{\boldsymbol{r}}{r}\, d\sigma \\ &= -D \int \nabla\rho_r \cdot \frac{\boldsymbol{r}}{r}\, r^2 \sin\theta\, d\theta d\phi \end{aligned} \tag{4.44}$$

with r, θ, ϕ the polar coordinates within the reference molecular frame and $d\sigma = r^2 \sin\theta\, d\theta d\phi$ the surface differential orthogonal to the radial vector $\boldsymbol{r}$ and hence orthogonal to the radial unit vector $\boldsymbol{r}/r$ ($\sin\theta\, d\theta d\phi$ being the solid angle differential). When considering reasonably diluted A molecules (i.e., any couple of A molecules can be considered at infinite distance), we can assume that for $r \geq r_0$ ρ_r be invariant over each spherical surface, that is $\rho_r(t, r)$ and thus

$$\frac{\partial\rho_r}{\partial x} = \frac{\partial\rho_r}{\partial r}\frac{x}{r} \tag{4.45}$$

$$\frac{\partial\rho_r}{\partial y} = \frac{\partial\rho_r}{\partial r}\frac{y}{r} \tag{4.46}$$

$$\frac{\partial\rho_r}{\partial z} = \frac{\partial\rho_r}{\partial r}\frac{z}{r} \tag{4.47}$$

Therefore, the B flux toward the A molecule at each radial distance $r \geq r_0$ can be expressed via

$$\begin{aligned}\Phi(t, r) &= -D \int \frac{\partial\rho_r}{\partial r}\frac{\boldsymbol{r}}{r} \cdot \frac{\boldsymbol{r}}{r}\, r^2 \sin\theta\, d\theta d\phi = -D \int \frac{\partial\rho_r}{\partial r}\, r^2 \sin\theta\, d\theta d\phi \\ &= -4\pi r^2 D \frac{\partial\rho_r}{\partial r}\end{aligned} \tag{4.48}$$

We can proceed further by assuming the steady state approximation for any spherical layer within the whole $r_0 \leq r < \infty$ range, that is $\Phi(t, r_0) \cong \Phi(t, r)$ providing

$$\frac{\partial\rho_r}{\partial r}(t, r) \cong -\frac{\Phi(t, r_0)}{4\pi r^2 D} \tag{4.49}$$

and hence

$$\begin{aligned}\rho_r(t, \infty) - \rho_r(t, r_0) &\cong -\int_{r_0}^{\infty} \frac{\Phi(t, r_0)}{4\pi r^2 D} dr = -\frac{\Phi(t, r_0)}{4\pi D}\left[-\frac{1}{r}\right]_{r_0}^{\infty} \\ &= -\frac{\Phi(t, r_0)}{4\pi D r_0}\end{aligned} \tag{4.50}$$

When realizing that $\rho_r(t, r_0) \cong 0$, $\rho_r(t, \infty) \cong [N_B]$ (i.e., the bulk molecular density $\rho_r(t, \infty)$ is virtually coinciding with the overall B molecular density $[N_B]$) and $[N_A]\Phi(t, r_0) \cong [\dot{N_B}]$ (with $[N_A]$ the overall A molecular density), from the reaction scheme Eqs. 4.40 and 4.50 we obtain

$$[\dot{N_B}] = [\dot{N_A}] = -[\dot{N_C}] \cong -4\pi D r_0 [N_B][N_A] \tag{4.51}$$

where obviously $[N_C]$ is the overall C molecular density.

Finally, when considering that the diffusion coefficient D corresponds to the B molecules diffusion relative to a reference A molecule (i.e., within the reference molecular frame), introducing the A and B Cartesian positions within the fixed laboratory reference frame (i.e., $\boldsymbol{r}_A$ and $\boldsymbol{r}_B$, respectively), we can express D via the A and B diffusion coefficients as obtained within the fixed laboratory reference frame (i.e., D_A and D_B)

$$\begin{aligned} 6Dt &= \langle|\boldsymbol{r}(t) - \boldsymbol{r}(0)|^2\rangle = \langle|\boldsymbol{r}_B(t) - \boldsymbol{r}_A(t) - [\boldsymbol{r}_B(0) - \boldsymbol{r}_A(0)]|^2\rangle \\ &= \langle|\Delta\boldsymbol{r}_B(t) - \Delta\boldsymbol{r}_A(t)|^2\rangle = \langle|\Delta\boldsymbol{r}_B(t)|^2 + |\Delta\boldsymbol{r}_A(t)|^2 - 2\Delta\boldsymbol{r}_B(t)\cdot\Delta\boldsymbol{r}_A(t)\rangle \\ &= \langle|\Delta\boldsymbol{r}_B(t)|^2\rangle + \langle|\Delta\boldsymbol{r}_A(t)|^2\rangle = 6\Big(D_B + D_A\Big)t \end{aligned} \tag{4.52}$$

with

$$\Delta\boldsymbol{r}_B(t) = \boldsymbol{r}_B(t) - \boldsymbol{r}_B(0) \tag{4.53}$$

$$\Delta\boldsymbol{r}_A(t) = \boldsymbol{r}_A(t) - \boldsymbol{r}_A(0) \tag{4.54}$$

and we used $\langle\Delta\boldsymbol{r}_B(t)\cdot\Delta\boldsymbol{r}_A(t)\rangle = 0$. From Eq. 4.52 we readily obtain $D = D_B + D_A$ and then Eq. 4.51 becomes (converting the molecular densities into molar densities, i.e., concentrations)

$$[\dot{B}] = [\dot{A}] = -[\dot{C}] \cong -4\pi\Big(D_B + D_A\Big)\mathcal{N}^0 r_0[A][B] = -k_A[A][B] \tag{4.55}$$

where $[A]$, $[B]$ and $[C]$ are the A, B and C overall concentrations, $\mathcal{N}^0$ is Avogadro's number and $k_A = 4\pi\Big(D_B + D_A\Big)\mathcal{N}^0 r_0$ provides the rate constant for the bimolecular reaction described by the scheme Eq. 4.40. It is worth noting that when $A = B$ and hence the reaction scheme is

$$2B \xrightarrow{k_A} C \tag{4.56}$$

Equation 4.55 reduces to

$$[\dot{B}] = -2[\dot{C}] \cong -16\pi D_B\mathcal{N}^0 r_0[B]^2 = -k_A[B]^2 \tag{4.57}$$

with thus $k_A = 16\pi D_B\mathcal{N}^0 r_0$.

4.3 Examples

4.3.1 Application to a Bimolecular Substitution Reaction in Solution

In this example we show how to model the bimolecular reaction $Cl^- + CH_3Br \rightarrow Br^- + CH_3Cl$ (see Fig. 4.2) in aqueous solution. This reaction (actually, this class of reactions termed as S_N2) has received a lot of attention and has represented, since the 1950s–60s, a sort of milestone for organic–physical chemistry studies [3] becoming also a didactically relevant paradigm of stereospecific reactions, present in (essentially) all the Organic Chemistry textbooks. More recently this reaction has also been at the center of a relevant number of theoretical–computational studies (see for example Refs. [4–6]). The reaction is known to consist (see Fig. 4.2) of a concerted process in which the two single bonds, involving the alkyl carbon atom, are simultaneously involved: one is formed and the other one is broken, with the transition structure (the transition state TS) involving a sort of hypervalent carbon atom. These features give this reaction the particularly high sensitivity to the polarity of the reaction medium [6]. Our aim here is to model the complete reaction process, reconstructing the overall rate constant to be compared to the experimental one at 25 °C in aqueous solution (3.2 10^{-6} M^{-1} s^{-1} [7]). The strategy adopted is based on the reaction scheme depicted in Fig. 4.3, where C_{NR} and C_R are the non-reactive and reactive complexes, respectively. C_{NR} indicates the chemical state where the two reactants (Cl^- and CH_3Br), although already interacting, are not yet close enough to alter their covalent structure and C_R is the chemical state corresponding to the RC potentially able to undergo to the covalent rearrangement.

From Eq. 4.55 we can express the association rate constant k_A via

$$k_A = 4\pi\left(D_{Cl^-} + D_{CH_3Br}\right)\mathcal{N}^0 r_0 \tag{4.58}$$

Fig. 4.2 Schematic view of the reaction under study

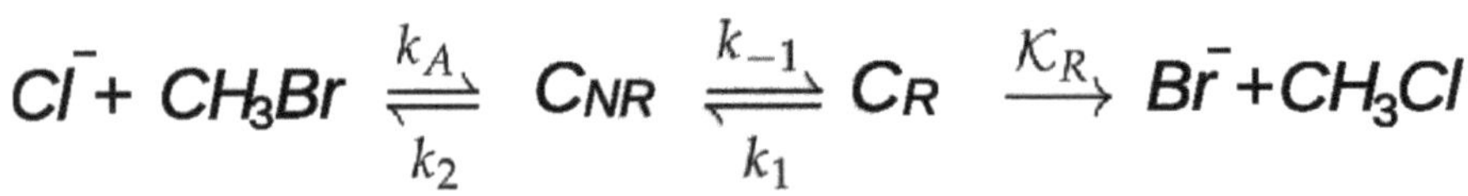

Fig. 4.3 Adopted reaction scheme

with $\mathcal{N}^0$ the Avogadro number, r_0 the Cl^- to CH_3Br radius length defining the overall reaction complex involving the interacting reactants (i.e., $C_{NR} + C_R$) and, finally, D_{Cl^-} and D_{CH_3Br} are the chloride ion and bromomethane diffusion coefficients, respectively. Moreover, $\mathcal{K}_R$ (see Eq. 4.28) is the rate constant involved in the covalent reactive event, k_2 is the rate constant for the C_{NR} dissociation providing the free diffusive reactants (i.e., the two reactants distance is larger than r_0) and k_1, k_{-1} are the rate constants for the $C_{NR} \rightleftharpoons C_R$ interconversion.

In order to model the kinetics of the reaction under study, making use of the theoretical background previously outlined, we addressed the following issues.

4.3.1.1 The Unperturbed Landau Free Energy Profile

First of all we have characterized the gas-phase (i.e., unperturbed) transformation of the $(Cl - CH_3 - Br)^-$ reactive complex (i.e., the C_R chemical state) toward the products by evaluating the electronic ground state energy variation as a function of the reaction coordinate ξ. Such unperturbed electronic energy profile was obtained within the framework of quantum chemical calculations [8], following a generally accepted procedure providing the Intrinsic Reaction Coordinate (IRC) [9] that we utilized as the proper reaction coordinate ξ. Briefly, the IRC procedure is accomplished by first locating the transition state structure, i.e., the electronic energy saddle point structure with mass weighted Hessian showing one eigenvector with negative eigenvalue (i.e., the eigenvector defining the IRC, basically the only C_R semiclassical internal coordinate) and then by constructing a set of 11 RC geometries as obtained by changing the IRC (i.e., the position along such eigenvector direction), minimizing the electronic ground eigenstate energy at each fixed IRC value (i.e., energy minimizing in the complementary orthogonal mass weighted subspace). Note that for typical reactions like the one we are considering, the energy minima along the IRC correspond to tiny shifts from the TS positions of the other mass weighted Hessian eigenvector coordinates, thus ensuring for the IRC the expected semiclassical behavior of the reaction coordinate ξ. The obtained electronic energies then provided the unperturbed Landau free energy profile along the reaction coordinate (see Fig. 4.4). It is worth to remark that, starting from the TS, we have spanned the reaction coordinate within the range where the atomic charges as obtained from quantum chemical calculations did not provide a full negative charge either confined on the Cl atom or on the Br atom, with only the extreme geometries basically corresponding to the Cl^- or Br^- conditions (the extreme left and right side ξ positions, respectively, in Fig. 4.4). Such a choice was motivated by considering that beyond such extreme positions no covalent rearrangement is present and the interacting chemical groups can be well described by classical mechanics.

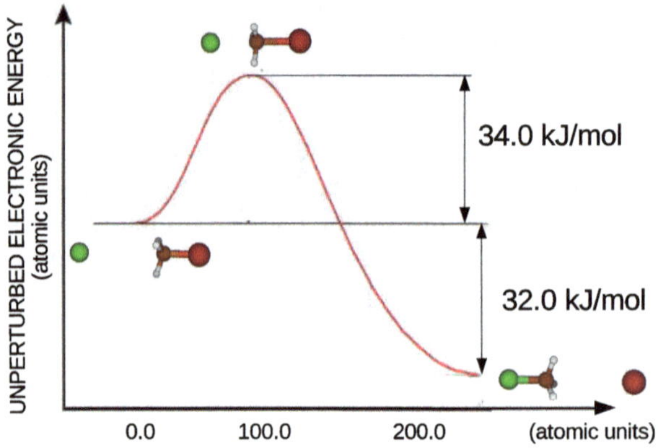

Fig. 4.4 Schematic view of the unperturbed Landau free energy profile given by the gas-phase electronic ground eigenstate energy profile as obtained by quantum chemical calculations. Note that the reaction coordinate origin coincides with the C_R maximum and C_{NR} minimum $Cl - CH_3$ distance, thus being located at the border of the two complexes

4.3.1.2 The Perturbed Landau Free Energy Profile

According to the first section of this chapter the rate constant $\mathcal{K}_R$ (see Eq. 4.28) of the covalent rearrangement reaction in solution we consider is entirely based on the determination of the mean TS traversing velocity, $\langle v \rangle = \sqrt{2k_BT/\pi}$ for a reaction coordinate corresponding to the IRC, and the Landau free energy profile of the solvated RC, with this latter evaluated through the theoretical–computational approach [10] already utilized in the calculation of the pKa reported at the end of Chap. 3. For such a procedure it is necessary to use the geometries as obtained for the unperturbed reaction (we assume as basically coincident the unperturbed and the perturbed reaction coordinates and related geometries), hereafter referred to as ξ_i geometries. Each of such C_R geometries was put at the center of a box with a fixed volume of 27 nm^3 filled with explicit water molecules at a density adjusted to reproduce the average pressure of a pure water box with identical volume and temperature at the experimental solvent density, i.e., for mimicking the experimental isobaric RC insertion. These boxes were then used for performing parallel MD simulations of 40 ns time length at 298 K. Note that each of these simulations was carried out using the (gas-phase) atomic charges of the $(Cl - CH_3 - Br)^-$ solute, kept frozen at the center of the box at a given ξ_i geometry (i.e., at fixed ξ_i position the RC is treated as a virtually rigid structure), thus providing the ξ_i geometry ensemble (we assume no significant solvent perturbation effects for the electronic ground eigenstate atomic charges). By using such MD simulations we obtained the perturbed Landau free energy change associated to the $\xi_i \rightarrow \xi_{i+1}$ step (virtually identical to the Helmholtz free energy change due to the RC transition from one δ interval centered in ξ_i to the δ interval centered in ξ_{i+1}, see Eqs. 4.33–4.34), by using Eq. 4.14 and reasonably assuming the RC mass tensor determinant as fully independent of the reaction coordinate [11]:

$$\begin{aligned}
\Delta\mathcal{A}_{\xi_i\to\xi_{i+1}} &= -k_BT\ln\frac{\mathcal{Q}(\xi_{i+1})}{\mathcal{Q}(\xi_i)} \cong -k_BT\ln\frac{\int e^{-\beta\mathcal{H}_0(\xi_{i+1},\zeta,\pi_\xi,\pi_\zeta)}d\zeta d\pi_\zeta d\pi_\xi}{\int e^{-\beta\mathcal{H}_0(\xi_i,\zeta,\pi_\xi,\pi_\zeta)}d\zeta d\pi_\zeta d\pi_\xi} \\
&= -k_BT\ln\frac{\int e^{-\beta\mathcal{H}_0(\xi_i,\zeta,\pi_\xi,\pi_\zeta)}\, e^{-\beta\Delta\mathcal{U}_{\xi_i\to\xi_{i+1}}}d\zeta d\pi_\zeta d\pi_\xi}{\int e^{-\beta\mathcal{H}_0(\xi_i,\zeta,\pi_\xi,\pi_\zeta)}d\zeta d\pi_\zeta d\pi_\xi} \\
&= -k_BT\ln\langle e^{-\beta\Delta\mathcal{U}_{\xi_i\to\xi_{i+1}}}\rangle_{\xi_i} \qquad (4.59)\\
\Delta\mathcal{U}_{\xi_i\to\xi_{i+1}} &= \mathcal{H}_0(\xi_{i+1},\zeta,\pi_\xi,\pi_\zeta)-\mathcal{H}_0(\xi_i,\zeta,\pi_\xi,\pi_\zeta) \qquad (4.60)
\end{aligned}$$

and hence for the Landau free energy change associated to the $\xi_i \to \xi_{i-1}$ step we can write

$$\begin{aligned}
\Delta\mathcal{A}_{\xi_i\to\xi_{i-1}} &\cong -k_BT\ln\langle e^{-\beta\Delta\mathcal{U}_{\xi_i\to\xi_{i-1}}}\rangle_{\xi_i} \qquad (4.61)\\
\Delta\mathcal{U}_{\xi_i\to\xi_{i-1}} &= \mathcal{H}_0(\xi_{i-1},\zeta,\pi_\xi,\pi_\zeta)-\mathcal{H}_0(\xi_i,\zeta,\pi_\xi,\pi_\zeta) \qquad (4.62)
\end{aligned}$$

with $\Delta\mathcal{U}_{\xi_i\to\xi_{i+1}}$ and $\Delta\mathcal{U}_{\xi_i\to\xi_{i-1}}$ the solution electronic ground eigenstate energy difference for the RC $\xi_i \to \xi_{i+1}$ and $\xi_i \to \xi_{i-1}$ step, respectively, and the ξ_i subscript of the angle brackets meaning that the ξ_i geometry ensemble was used (when considering a reaction as occurring in an excited electronic eigenstate the same energy changes clearly refer to the excited electronic eigenstate energy). Note that once assuming the RC mass tensor determinant as independent of the reaction coordinate, the MD simulations with frozen $\xi = \xi_i$ geometries are unaffected by the Fixman theorem [12] and thus can be used to obtain the proper statistical ensembles. Repeating such a procedure for all the 11 ξ_i geometry MD simulations, we could reconstruct the whole perturbed Landau free energy profile taken as the average between the Landau free energy curves as obtained by the $\xi_i \to \xi_{i+1}$ and the $\xi_i \to \xi_{i-1}$ steps (the results are reported in Fig. 4.5). As expected [6] Fig. 4.5 clearly shows that the presence of the polar solvent largely increases the $R \to TS$ free energy barrier from 34 to 100 kJ/mol, well matching the experimental estimate (99 kJ/mole [13]) and providing, according to Eq. 4.28, $\mathcal{K}_R = 2.0\ 10^{-5}\ s^{-1}$.

4.3.1.3 The Association k_A and Dissociation k_2 Rate Constants

As previously reported, for evaluating the C_{NR} formation rate constant k_A (see Eq. 4.58) we need r_0, essentially corresponding to the threshold beyond which we can consider free diffusive reactants, as well as the reactants (aqueous solution) diffusion coefficients. For the latter we have used the experimental values, as reported in [14, 15], equal to 2.0 10^{-9} m^2 s^{-1} and 1.5 10^{-9} m^2 s^{-1} for chloride ion and bromomethane, respectively. For the evaluation of r_0 we performed a further MD simulation of 100 ns at 298 K, hereafter termed as MD-free simulation, in which Cl^- and CH_3Br are simulated as classical freely moving species in a box with fixed volume as large as 216 nm^3 filled by water molecules again at the proper density (i.e., mimicking the isobaric insertion of the solutes). From such MD-free simulation, fully sampling the C_{NR} chemical state and accessing the most stable configurations

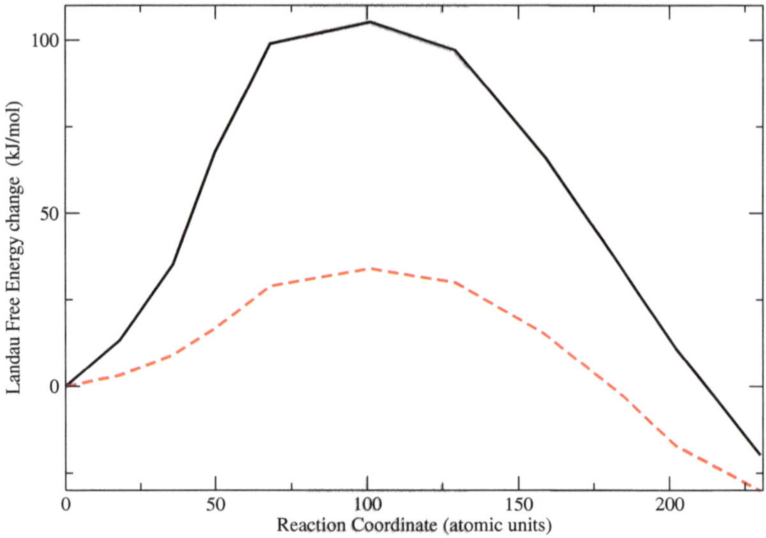

Fig. 4.5 Landau free energy profile in aqueous solution (black solid line) and in the gas-phase (red dotted line, see Fig. 4.4)

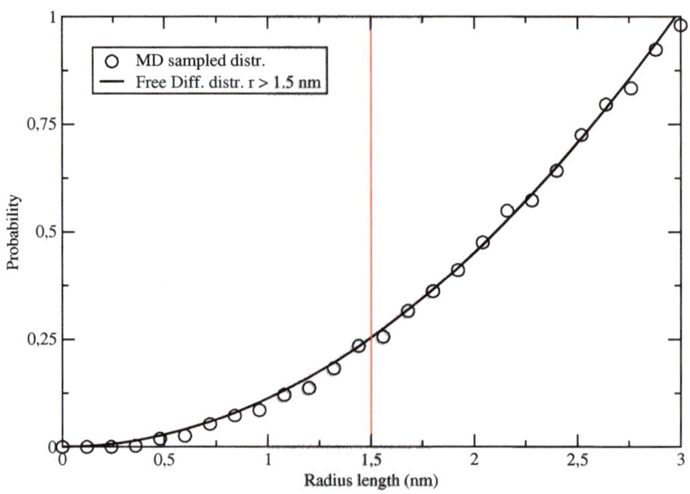

Fig. 4.6 Probability density of the distance between chloride ion and CH_3Br, as obtained by the MD-free simulation (circles), and the analytical quadratic curve providing the free diffusive behavior (black solid line). The red vertical solid line indicates the r_0 radius

of the C_R chemical state (i.e., $\xi \approx 0.0$), we evaluated the probability distribution of the Cl^- to CH_3Br distance, as reported in Fig. 4.6. The r_0 value was then estimated by the minimal distance where the probability distribution begins to exhibit a proper quadratic shape, i.e., indicating for the two reactants a free diffusive behavior with

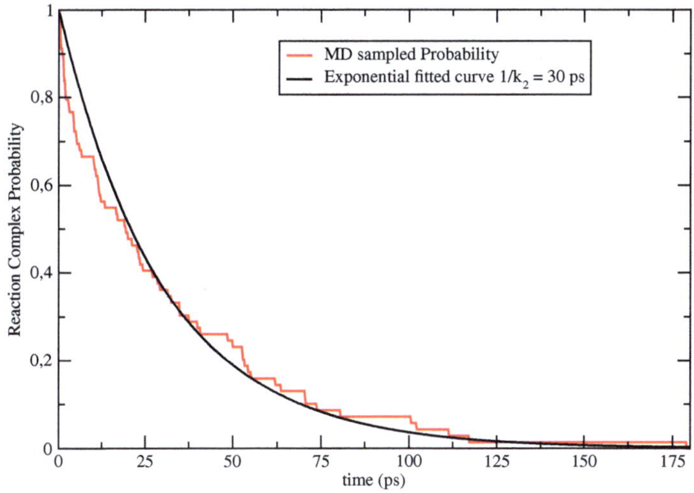

Fig. 4.7 Time decay of the $C_{NR} + C_R$ overall reaction complex as provided by the MD-free simulation (red solid line) and the corresponding exponential fitting curve (black solid line)

hence no C_{NR} chemical state formed. The obtained value (1.5 nm) is in line with similar calculations in other systems previously investigated [16]. Combining the obtained r_0 with the experimental diffusion coefficients according to Eq. 4.58, provided $k_A = 4.0\ 10^{10}\ \mathrm{M}^{-1}\ \mathrm{s}^{-1}$. The estimate of k_2, i.e., the rate constant for the C_{NR} dissociation to the free reactants, was computed by using the MD-free simulation to obtain several subtrajectories, each starting within the r_0 sphere, and then monitoring the time decay of the fraction corresponding to the Cl^- to CH_3Br distances still within the r_0 limit (i.e., the time decay of the overall reaction complex). The results, reported in Fig. 4.7 and as expected showing a first-order decay according to the likely pre-equilibrium condition of the $[C_R]/[C_{NR}]$ ratio [16], were fitted by a monoexponential function providing $k_2 = 3.33\ 10^{10}\ \mathrm{s}^{-1}$.

4.3.1.4 The Overall S_N2 Rate Constant

Finally, considering that $\mathcal{K}_R \ll k_A$ and hence we can extend the pre-equilibrium condition also to the $\frac{[C_{NR}]}{[Cl^-][CH_3Br]}$ ratio, allowed us to express the overall rate constant K_{S_N2} via [16]

$$K_{S_N2} \cong \frac{k_A}{k_2}\frac{k_{-1}}{k_1}\mathcal{K}_R \tag{4.63}$$

with k_{-1}/k_1 as obtained by the MD-free simulation employing the projections of the mass weighted Cartesian coordinates of the Cl^- and CH_3Br reactants over the IRC eigenvector to identify the C_R and C_{NR} configurations and thus their equilibrium ratio $k_{-1}/k_1 \cong 0.1$ at 298 K. The use in Eq. 4.63 of the obtained $\mathcal{K}_R$, k_A, k_2 and k_{-1}/k_1

values then provided $K_{S_N2} = 2.4\ 10^{-6}\ M^{-1}\ s^{-1}$, well matching the experimentally measured rate constant ($3.2\ 10^{-6}\ M^{-1}\ s^{-1}$ [7]). Interestingly, by adopting the Eyring approximation (see Eq. 4.32) instead of Eq. 4.28 for evaluating $\mathcal{K}_R$, we obtained $K_{S_N2} = 2.7\ 10^{-7}\ M^{-1}\ s^{-1}$ clearly showing the inaccuracy of such an approximation for the S_N2 reaction considered.

References

1. Landau L (1937) On the theory of phase transitions. Zh Eksp Teor Fiz 7:19
2. Hammes GG (1978) Principles of chemical kinetics. Academic, New York
3. March J (1985) Advanced organic chemistry: reactions, mechanisms, and structure. Wiley, New York
4. Chandrasekhar J, Smith SF, Jorgensen WL (1984) SN2 reaction profiles in the gas phase and aqueous solution. J Am Chem Soc 106:3049
5. Raugei S, Cardini G, Schettino V (2001) Microsolvation effect on chemical reactivity: the case of the Cl^- + CH_3Br SN2 reaction. J Chem Phys 114:4089
6. Valverde D, Georg HC, Canuto S (2022) Free-energy landscape of the SN2 reaction CH_3Br + $Cl^- \rightarrow CH_3Cl + Br^-$ in different liquid environments. J Phys Chem B 126:3685
7. Alexander R, Ko ECF, Parker AJ, Broxton TJ (1968) Solvation of ions. XIV. Protic-dipolar aprotic solvent effects on rates of bimolecular reactions. Solvent activity coefficients of reactants and transition states at 25°. J Am Chem Soc 90:5049
8. The IRC calculations were performed using the cam-b3lyp functional in conjunction with the 6-31+G* basis set. Single pont calculations were subsequently carried out on the points along the coordinate with the Coupled Cluster method, i.e. ccsd(t), in conjunction with the 6-311+G** atomic basis set
9. Fukui K. (1981) The path of chemical reactions - The IRC approach. Acc Chem Res 14:363
10. Amadei A, D'Alessandro M, D'Abramo M, Aschi M (2009) Theoretical characterization of electronic states in interacting chemical systems. J Chem Phys 130:084109
11. We always consider the vibrational partition function and the RC mass tensor determinant as independent of the reaction coordinate and thus irrelevant for the Landau free energy change along ξ (see also Chapter 3 for further discussions on this specific point)
12. Amadei A, Chillemi G, Ceruso MA, Grottesi A, Di Nola A (2000) Molecular dynamics simulations with constrained roto-translational motions: theoretical basis and statistical mechanical consistency. J Chem Phys 112:9–23
13. Bathgate RH, Moelwyn-Hughes EA (1959) The kinetics of certain ionic exchange reactions of the four methyl halides in aqueous solution. J Chem Soc 2642–2648
14. Hashimoto K, Otsuki N, Saito T, Yokota H (2023) Application of electrical treatment to alteration of cementitious material due to leaching. Molecules 28:2152
15. De Bruyn WJ, Saltzman ES (1997) Modelling complex bimolecular reactions in a condensed phase: the case of phosphodiester hydrolysis. Marine Chem 57:55
16. Nardi AN, Olivieri A, D'Abramo M, Amadei A (2024) A theoretical-computational study of phosphodiester bond cleavage kinetics as a function of the temperature. ChemPhysChem 25:e202300952

Chapter 5
Appendix: Physical States and Observables in Quantum Systems

Abstract In this Appendix we provide an overview of the basic principles and derivations of quantum mechanics, essential for a statistical mechanical treatment of complex molecular systems.

5.1 The Dirac Generalized Vector Space

For classical mechanical systems the knowledge of the positions and momenta at any instant of time is a sufficient condition for accessing whatever observable of the system (i.e., for each microscopic physical state we can exactly define the corresponding value of any mechanical property). Unfortunately, for quantum mechanical systems it is not possible to know at a given instant of time the exact positions and momenta of the particles. Hence, a change of viewpoint is necessary for treating physical–chemical systems at quantum level. Paul Dirac introduced a powerful and very general formalism [1] to define quantum states and their relation with the system observables. Within such an approach, the state space of a mechanical system is conceived as a vector space, thus implying that each possible quantum state can be expressed as a linear combination of the reference vectors defining the basis set of the vector space. However, such a Dirac's vector space is indeed a generalization of the usual vector spaces as (i) the basis set is in principle always of infinite dimension (like in the Hilbert space) and (ii) the reference vectors providing the basis set are not necessarily of finite length. Any quantum state is actually given by a generalized vector which can be expressed in two different forms: the ket $|\ \rangle$ form and the bra $\langle\ |$ form. Therefore, we can identify a generic "a" quantum state via its ket $|a\rangle$ or bra $\langle a|$ form, which can be used to obtain the square length of such a vector by means of their product (the scalar or Hermitian product in Dirac's space) indicated by $\langle a|a\rangle$ and always providing a real positive number (i.e., $\langle a|a\rangle \geq 0$). Note that given two quantum states $|a\rangle$ and $|b\rangle$ we can obtain their product into two ways: $\langle a|b\rangle$ or $\langle b|a\rangle$ which provide a scalar (in general a complex number) where $\langle a|b\rangle = \langle b|a\rangle^*$, i.e., $\langle a|b\rangle$ is the complex conjugate of $\langle b|a\rangle$, and hence $\langle a|$ is often defined as the imaginary conjugate of $|a\rangle$. Different types of (generalized) vectors can be used to define a proper basis set: (i) an infinite set of independent vectors identified by an

A. Amadei and M. Aschi, *Statistical Mechanics for Chemical Thermodynamics and Kinetics*, https://doi.org/10.1007/978-3-031-77929-9_5

infinite set of n discrete variables $\boldsymbol{\lambda}_l = (\lambda_{1,l}\ \lambda_{2,l}\ \dots\ \lambda_{n,l})^T$ with $l = 1, \infty$ (discrete basis set), (ii) an infinite set of independent vectors identified by m continuous variables $\boldsymbol{x} = (x_1\ x_2\ \dots\ x_m)^T$ (continuous basis set) and in the most general case (iii) an infinite set of independent vectors identified by a combination of discrete and continuous variables (mixed basis set). Therefore, the arbitrary ket quantum state $|a\rangle$, according to the type of basis set which is used, can be expressed as

$$\textit{discrete basis set}$$
$$|a\rangle = \sum_l a_l\ |\eta_l\rangle \tag{5.1}$$
$$\textit{continuous basis set}$$
$$|a\rangle = \int a(\boldsymbol{x})\ |\eta(\boldsymbol{x})\rangle\ d\boldsymbol{x} + \sum_k c_{\boldsymbol{\chi}_k}\ |\eta(\boldsymbol{\chi}_k)\rangle \tag{5.2}$$
$$\textit{mixed basis set}$$
$$|a\rangle = \sum_l \int a_l(\boldsymbol{x})\ |\eta_l(\boldsymbol{x})\rangle\ d\boldsymbol{x} + \sum_l \sum_k c_{l,\boldsymbol{\chi}_k}\ |\eta_l(\boldsymbol{\chi}_k)\rangle \tag{5.3}$$

where $|\eta_l\rangle$, $|\eta(\boldsymbol{x})\rangle$, $|\eta_l(\boldsymbol{x})\rangle$ are, respectively, the discrete, continuous and mixed basis set vectors, $a_l, a(\boldsymbol{x}), a_l(\boldsymbol{x})$ the corresponding coefficients (in general complex numbers) involved in the linear combination and in Eqs. 5.2 and 5.3 $\boldsymbol{\chi}_k$ is the generic kth value of a discrete set of the continuous $\boldsymbol{x}$ vector with hence $\sum_k$ (similarly to $\sum_l$) indicating the summation over such discrete values, with $c_{\boldsymbol{\chi}_k}, c_{l,\boldsymbol{\chi}_k}$ the corresponding coefficients. Note that $a(\boldsymbol{x}), a_l(\boldsymbol{x})$ must be continuous functions of $\boldsymbol{x}$ and hence the summation over the discrete $\boldsymbol{\chi}_k$ values in Eqs. 5.2 and 5.3 is introduced to include terms due to non-continuous reference vectors and coefficients providing linear combinations which cannot be obtained by integration.

It is easy to express also the bra quantum state $\langle a|$ from the previous equations, considering that the bra is the imaginary conjugate of the ket

$$\textit{discrete basis set}$$
$$\langle a| = \sum_l a_l^*\ \langle \eta_l| \tag{5.4}$$
$$\textit{continuous basis set}$$
$$\langle a| = \int a^*(\boldsymbol{x})\ \langle \eta(\boldsymbol{x})|\ d\boldsymbol{x} + \sum_k (c_{\boldsymbol{\chi}_k})^*\ \langle \eta(\boldsymbol{\chi}_k)| \tag{5.5}$$
$$\textit{mixed basis set}$$
$$\langle a| = \sum_l \int a_l^*(\boldsymbol{x})\ \langle \eta_l(\boldsymbol{x})|\ d\boldsymbol{x} + \sum_l \sum_k (c_{l,\boldsymbol{\chi}_k})^*\ \langle \eta_l(\boldsymbol{\chi}_k)| \tag{5.6}$$

It must be remarked that when using an orthonormal discrete basis set, i.e., $\langle \eta_l|\eta_{l'}\rangle = \delta_{l,l'}$ with $\delta_{l,l'}$ Kronecker's delta for the l index of the discrete variables $\boldsymbol{\lambda}_l$, any a_l coefficient is simply the projection of the quantum state $|a\rangle$ onto the reference vector $|\eta_l\rangle$, that is (see Eq. 5.1)

$$\langle \eta_{l'} | a \rangle = \sum_{l} a_l \langle \eta_{l'} | \eta_l \rangle = a_{l'} \tag{5.7}$$

Equation 5.7 obviously implies that in the $|\eta_l\rangle$ basis set we can express $|a\rangle$ as a column vector with the components given by the a_l coefficients

$$|a\rangle = \boldsymbol{a} = \begin{bmatrix} a_1 \\ a_2 \\ \cdot \\ a_n \\ \cdot \end{bmatrix} \tag{5.8}$$

Therefore, in such a basis set we must necessarily have $\langle a| = (\boldsymbol{a}^*)^T = (\boldsymbol{a}^T)^*$, as it follows from the conjugate complex of Eq. 5.7 and, more in general, clearly shown by the Hermitian product of two quantum states $|a\rangle$ and $|b\rangle$ both expressed via the discrete orthonormal basis set $|\eta_l\rangle$

$$\langle a|b\rangle = \sum_{l}\sum_{l'} a_l^* b_{l'} \langle \eta_l | \eta_{l'} \rangle = \sum_{l} a_l^* b_l = (\boldsymbol{a}^*)^T \boldsymbol{b} \tag{5.9}$$

providing

$$\langle a|a\rangle = (\boldsymbol{a}^*)^T \boldsymbol{a} = \sum_{l} |a_l|^2 \geq 0 \tag{5.10}$$

with obviously $\boldsymbol{b}$ the column vector expressing $|b\rangle$ within the $|\eta_l\rangle$ basis set. Equations 5.9 and 5.10 plainly show that when we consider an orthonormal discrete basis set the Dirac generalized space is equivalent to the Hilbert space and hence the reference ket or bra vectors defining the basis set must correspond to a set of reference quantum states which can be expressed by the column $\boldsymbol{\eta}_l$ (ket form) or row $(\boldsymbol{\eta}_l^*)^T$ (bra form) orthogonal unit vectors (i.e., with components $[\boldsymbol{\eta}_l]_{l'} = \delta_{l,l'}$ when they are expressed within their own basis set).

The peculiar feature of the Dirac generalized vector space is actually the possibility to consider basis set vectors as defined by continuous variables, thus requiring a specific definition of orthogonality well corresponding to the one used for a discrete basis set. Therefore, when defining the orthogonality of such basis set vectors via

$$\langle \eta(\boldsymbol{x}_1) | \eta(\boldsymbol{x}_2) \rangle = \delta(\boldsymbol{x}_1 - \boldsymbol{x}_2) = \delta(\boldsymbol{x}_2 - \boldsymbol{x}_1) \tag{5.11}$$

$$\langle \eta_{l_1}(\boldsymbol{x}_1) | \eta_{l_2}(\boldsymbol{x}_2) \rangle = \delta_{l_1,l_2} \delta(\boldsymbol{x}_1 - \boldsymbol{x}_2) = \delta_{l_1,l_2} \delta(\boldsymbol{x}_2 - \boldsymbol{x}_1) \tag{5.12}$$

with $\delta(x)$ the Dirac function, from Eqs. 5.2 and 5.3 and considering $\boldsymbol{x}_1 = (x_{1,1}\ x_{2,1}\ \ldots\ x_{m,1})^T$ not included into the $\boldsymbol{\chi}_k$ discrete set or simply $c_{\chi_k} = c_{l,\chi_k} = 0$, we obtain for the continuous and mixed basis set, respectively,

$$\langle \eta(\boldsymbol{x}_1) | a \rangle = \int a(\boldsymbol{x})\ \langle \eta(\boldsymbol{x}_1) | \eta(\boldsymbol{x}) \rangle\ d\boldsymbol{x} = \int a(\boldsymbol{x})\ \delta(\boldsymbol{x} - \boldsymbol{x}_1)\ d\boldsymbol{x} = a(\boldsymbol{x}_1) \tag{5.13}$$

$$\langle \eta_{l_1}(\boldsymbol{x}_1)|a\rangle = \sum_l \int a_l(\boldsymbol{x}) \, \langle \eta_{l_1}(\boldsymbol{x}_1)|\eta_l(\boldsymbol{x})\rangle \, d\boldsymbol{x} = \sum_l \delta_{l,l_1} \int a_l(\boldsymbol{x}) \, \delta(\boldsymbol{x} - \boldsymbol{x}_1) \, d\boldsymbol{x}$$
$$= a_{l_1}(\boldsymbol{x}_1) \tag{5.14}$$

It is worth to note that the orthogonality definition given in Eqs. 5.11 and 5.12 implies that each reference vector of the continuous or mixed basis set has infinite length, as it follows from

$$\lim_{\boldsymbol{x}_2 \to \boldsymbol{x}_1} \langle \eta(\boldsymbol{x}_1)|\eta(\boldsymbol{x}_2)\rangle = \lim_{\boldsymbol{x}_2 \to \boldsymbol{x}_1} \delta(\boldsymbol{x}_1 - \boldsymbol{x}_2) = \infty \tag{5.15}$$
$$\lim_{\boldsymbol{x}_2 \to \boldsymbol{x}_1} \langle \eta_{l_1}(\boldsymbol{x}_1)|\eta_{l_1}(\boldsymbol{x}_2)\rangle = \lim_{\boldsymbol{x}_2 \to \boldsymbol{x}_1} \delta(\boldsymbol{x}_1 - \boldsymbol{x}_2) = \infty \tag{5.16}$$

Equations 5.13 and 5.14 clearly show that for a quantum state $|a\rangle$ not involving the summation over the discrete subset χ_k, i.e., a quantum state defined by a finite length vector with $c_{\chi_k} = 0$ and $c_{l,\chi_k} = 0$, the continuous function $a(\boldsymbol{x})$ or the column vector $\boldsymbol{a}(\boldsymbol{x})$ with components given by the continuous functions $a_l(\boldsymbol{x})$ express such a quantum state in the $|\eta(\boldsymbol{x})\rangle$ or $|\eta_l(\boldsymbol{x})\rangle$ basis set, respectively, i.e., $|a\rangle = a(\boldsymbol{x})$ or $|a\rangle = \boldsymbol{a}(\boldsymbol{x})$ (note that $a(\boldsymbol{x})$ and $\boldsymbol{a}(\boldsymbol{x})$ are usually termed as wavefunctions). These last derivations readily provide, when using either the continuous or the mixed basis set to express any of the corresponding reference vectors, the following relations (see Eqs. 5.11 and 5.12):

$$|\eta(\boldsymbol{x}_1)\rangle = \delta(\boldsymbol{x} - \boldsymbol{x}_1) \tag{5.17}$$
$$\langle \eta(\boldsymbol{x}_1)| = \delta(\boldsymbol{x} - \boldsymbol{x}_1) \tag{5.18}$$
$$|\eta_{l_1}(\boldsymbol{x}_1)\rangle = \boldsymbol{\delta}_{l_1} \delta(\boldsymbol{x} - \boldsymbol{x}_1) \tag{5.19}$$
$$\langle \eta_{l_1}(\boldsymbol{x}_1)| = (\boldsymbol{\delta}_{l_1})^T \delta(\boldsymbol{x} - \boldsymbol{x}_1) \tag{5.20}$$

where $\boldsymbol{\delta}_{l_1}$ is the column unit vector with the l component given by δ_{l,l_1}. More in general we can define the mixed basis set reference vectors by expressing $\boldsymbol{\delta}_{l_1}$ via any arbitrary orthonormal basis set, providing

$$|\eta_{l_1}(\boldsymbol{x}_1)\rangle = \boldsymbol{\eta}_{l_1} \delta(\boldsymbol{x} - \boldsymbol{x}_1) \tag{5.21}$$
$$\langle \eta_{l_1}(\boldsymbol{x}_1)| = (\boldsymbol{\eta}_{l_1}^*)^T \delta(\boldsymbol{x} - \boldsymbol{x}_1) \tag{5.22}$$

with $\boldsymbol{\eta}_{l_1}$ the column unit vector expressing $\boldsymbol{\delta}_{l_1}$ into the new basis set. Hence, from the equivalence between column vectors and ket generalized vectors within a space defined by a discrete basis set, we obtain

$$|\eta_{l_1}(\boldsymbol{x}_1)\rangle = \delta(\boldsymbol{x} - \boldsymbol{x}_1) \, |\eta_{l_1}\rangle \tag{5.23}$$
$$\langle \eta_{l_1}(\boldsymbol{x}_1)| = \delta(\boldsymbol{x} - \boldsymbol{x}_1) \, \langle \eta_{l_1}| \tag{5.24}$$

where the last two equations provide a very general expression of the mixed basis set reference vectors with their parts associated to the discrete variables expressed as ket (bra) generalized vectors.

Finally, using Eqs. 5.13 and 5.14 and the definition of the Hermitian product within a continuous or mixed (orthogonal) basis set we can express the product of the finite length (i.e., $c_{\chi_k} = 0$ and $c_{l,\chi_k} = 0$) $|a\rangle$ and $|b\rangle$ vectors as

$$\begin{aligned}
& \textit{continuous basis set} \\
\langle a|b\rangle &= \int a^*(\boldsymbol{x})\ \langle \eta(\boldsymbol{x}|\eta(\boldsymbol{x}')\rangle\ b(\boldsymbol{x}')\ d\boldsymbol{x}d\boldsymbol{x}' = \int a^*(\boldsymbol{x})b(\boldsymbol{x}')\ \delta(\boldsymbol{x}' - \boldsymbol{x})\ d\boldsymbol{x}d\boldsymbol{x}' \\
&= \int a^*(\boldsymbol{x})b(\boldsymbol{x})d\boldsymbol{x} \qquad (5.25) \\
& \textit{mixed basis set} \\
\langle a|b\rangle &= \sum_l \sum_{l'} \int a_l^*(\boldsymbol{x})\ \langle \eta_l(\boldsymbol{x}|\eta_{l'}(\boldsymbol{x}')\rangle\ b_{l'}(\boldsymbol{x}')\ d\boldsymbol{x}d\boldsymbol{x}' \\
&= \sum_l \int a_l^*(\boldsymbol{x})\ \langle \eta_l(\boldsymbol{x}|\eta_l(\boldsymbol{x}')\rangle\ b_l(\boldsymbol{x}')\ d\boldsymbol{x}d\boldsymbol{x}' \\
&= \sum_l \int a_l^*(\boldsymbol{x})b_l(\boldsymbol{x}')\ \delta(\boldsymbol{x}' - \boldsymbol{x})\ d\boldsymbol{x}d\boldsymbol{x}' = \sum_l \int a_l^*(\boldsymbol{x})b_l(\boldsymbol{x})d\boldsymbol{x} \\
&= \int \sum_l a_l^*(\boldsymbol{x})b_l(\boldsymbol{x})d\boldsymbol{x} = \int (\boldsymbol{a}^*(\boldsymbol{x}))^T \boldsymbol{b}(\boldsymbol{x})d\boldsymbol{x} \qquad (5.26)
\end{aligned}$$

where obviously $b(\boldsymbol{x})$ and $\boldsymbol{b}(\boldsymbol{x})$ are the wavefunctions expressing $|b\rangle$ in the $|\eta(\boldsymbol{x})\rangle$ and $|\eta_l(\boldsymbol{x})\rangle$ basis set, respectively.

5.2 Observables and Operators

One of the main ideas underlying quantum mechanics is that any mechanical property (observable) α can be represented by a linear operator $\widehat{\alpha}$, acting on any vector (state) to provide a new vector (new state)

$$\widehat{\alpha}\ |a\rangle = |b\rangle \qquad (5.27)$$

Linear operators are characterized by the following general properties:

$$\begin{aligned}
\widehat{\alpha} c|a\rangle &= c\widehat{\alpha}|a\rangle \qquad (5.28) \\
\widehat{\alpha}\left(|a\rangle + |a'\rangle\right) &= \widehat{\alpha}|a\rangle + \widehat{\alpha}|a'\rangle \qquad (5.29) \\
\left(\widehat{\beta}\widehat{\alpha}\right)|a\rangle &= \widehat{\beta}\left(\widehat{\alpha}|a\rangle\right) \qquad (5.30)
\end{aligned}$$

where c is a scalar and $\widehat{\beta}$ another linear operator (note that usually $\widehat{\beta}\widehat{\alpha} \neq \widehat{\alpha}\widehat{\beta}$). We can now introduce the inverse operator $\widehat{\alpha}^{-1}$ as defined by

$$\widehat{\alpha}^{-1}\widehat{\alpha} = \widehat{\alpha}\widehat{\alpha}^{-1} = \widehat{I} \tag{5.31}$$

($\widehat{I}$ is the identity operator) and the adjoint operator $\widehat{\alpha}^{\dagger}$ defined via the bra form of Eq. 5.27

$$\langle b| = \langle \widehat{\alpha} a| = \langle a|\widehat{\alpha}^{\dagger} \tag{5.32}$$

Equations 5.31 and 5.32 also imply

$$(\widehat{\alpha}\widehat{\beta})^{-1} = \widehat{\beta}^{-1}\widehat{\alpha}^{-1} \tag{5.33}$$

$$\begin{aligned} \langle \widehat{\alpha}\widehat{\beta} a| &= \langle a|(\widehat{\alpha}\widehat{\beta})^{\dagger} = \langle \widehat{\alpha}(\widehat{\beta} a)| \\ &= \langle \widehat{\beta} a|\widehat{\alpha}^{\dagger} = \langle a|\widehat{\beta}^{\dagger}\widehat{\alpha}^{\dagger} \end{aligned} \tag{5.34}$$

readily providing

$$(\widehat{\alpha}\widehat{\beta})^{\dagger} = \widehat{\beta}^{\dagger}\widehat{\alpha}^{\dagger} \tag{5.35}$$

Note that the linear operator $\widehat{\alpha}^{\dagger}$ (the adjoint of $\widehat{\alpha}$) as defined by Eq. 5.32 plays the analogous role for operators of the complex conjugate for scalars. Moreover, the operator $\widehat{\alpha}$ is defined as unitary operator if $\widehat{\alpha}\widehat{\alpha}^{\dagger} = \widehat{\alpha}^{\dagger}\widehat{\alpha} = \widehat{I}$, clearly meaning that $\widehat{\alpha}^{\dagger} = \widehat{\alpha}^{-1}$, and as Hermitian operator if $\widehat{\alpha} = \widehat{\alpha}^{\dagger}$. In case $\widehat{\alpha}$ is unitary and Hermitian at the same time, then necessarily $\widehat{\alpha} = \widehat{I}$. From the definition of the Hermitian product (the scalar product in Dirac space) we can write $\langle b|a'\rangle = \langle a'|b\rangle^*$ and hence by using Eqs. 5.27 and 5.32 we obtain

$$\langle a|\widehat{\alpha}^{\dagger}|a'\rangle = \langle a'|\widehat{\alpha}|a\rangle^* \tag{5.36}$$

By using orthonormal discrete basis set vectors we can also define the projection operator $|\eta_{l'}\rangle\langle\eta_{l'}|$ providing, when acting on the quantum state $|a\rangle$, the projection vector of $|a\rangle$ onto the reference ket vector $|\eta_{l'}\rangle$, as it follows from (see Eq. 5.1)

$$|\eta_{l'}\rangle\langle\eta_{l'}|\ |a\rangle = \sum_{l} a_l |\eta_{l'}\rangle\langle\eta_{l'}|\eta_l\rangle = a_{l'}|\eta_{l'}\rangle \tag{5.37}$$

The last equation implies that $\sum_{l'} |\eta_{l'}\rangle\langle\eta_{l'}| = \widehat{I}$, furnishing a very general definition of the identity operator which can be used to express any linear operator within the discrete basis set $|\eta_l\rangle$. Note that Eq. 5.37 and hence the expression of the identity operator by means of the complete basis set can be easily extended to a continuous or mixed basis set providing fully equivalent expressions, although clearly involving the integral over the continuous variables defining the basis set vectors. From the definition of the identity operator, i.e., $\widehat{I}|b\rangle = |b\rangle$, we can write (see Eqs. 5.1 and 5.27)

$$|b\rangle = \widehat{I}\widehat{\alpha}\sum_l a_l|\eta_l\rangle = \sum_{l'}\sum_l \langle\eta_{l'}|\widehat{\alpha}|\eta_l\rangle a_l|\eta_{l'}\rangle \tag{5.38}$$

where, from the previous section, each $|\eta_{l'}\rangle$ reference ket can be expressed, within its own basis set, by a unit column vector $\boldsymbol{\eta}_{l'}$ with components given by $\delta_{l,l'}$. Therefore, from Eq. 5.38 we have

$$|b\rangle = \widehat{\alpha}|a\rangle = \widetilde{\alpha}\boldsymbol{a} = \boldsymbol{b} \tag{5.39}$$

with $\boldsymbol{a}$ the column vector expressing $|a\rangle$ in the $|\eta_l\rangle$ basis set (i.e., given by the a_l coefficients), $\widetilde{\alpha}$ a matrix (in principle an infinite dimensional matrix) whose l', l element is given by

$$[\widetilde{\alpha}]_{l',l} = \langle\eta_{l'}|\widehat{\alpha}|\eta_l\rangle \tag{5.40}$$

and $\boldsymbol{b}$ the column vector expressing $|b\rangle$ in the same basis set. The definition of the matrix elements representing a linear operator within the discrete basis set can also be used, in conjunction with Eq. 5.36, to obtain

$$\widetilde{\alpha}^\dagger = (\widetilde{\alpha}^*)^T \tag{5.41}$$

where $\widetilde{\alpha}^\dagger$ (the adjoint of $\widetilde{\alpha}$) is obviously the matrix representing $\widehat{\alpha}^\dagger$ within the $|\eta_l\rangle$ basis set. By using the last derivations we can then write

$$|\eta_{l'}\rangle\langle\eta_{l'}| = \boldsymbol{\eta}_{l'}\left(\boldsymbol{\eta}_{l'}^*\right)^T \tag{5.42}$$

$$\sum_{l'}|\eta_{l'}\rangle\langle\eta_{l'}| = \sum_{l'}\boldsymbol{\eta}_{l'}\left(\boldsymbol{\eta}_{l'}^*\right)^T = \widetilde{I} \tag{5.43}$$

with $\widetilde{I}$ the identity matrix (note that these last two equations are valid for whatever orthonormal basis set used to express the $\boldsymbol{\eta}_{l'}$ unit vectors). Finally, from the definition of the identity operator and of the elements of the matrix expressing an operator within the orthonormal discrete basis set we obtain, once defined $\widehat{\gamma} = \widehat{\alpha}\widehat{\beta}$,

$$\begin{aligned}[\widetilde{\gamma}]_{l',l} &= \langle\eta_{l'}|\widehat{\alpha}\widehat{\beta}|\eta_l\rangle = \langle\eta_{l'}|\widehat{\alpha}\left(\sum_{l''}|\eta_{l''}\rangle\langle\eta_{l''}|\right)\widehat{\beta}|\eta_l\rangle \\ &= \sum_{l''}\langle\eta_{l'}|\widehat{\alpha}|\eta_{l''}\rangle\langle\eta_{l''}|\widehat{\beta}|\eta_l\rangle\end{aligned} \tag{5.44}$$

thus implying $\widetilde{\gamma} = \widetilde{\alpha}\widetilde{\beta}$.

5.3 Eigenstates and Eigenvalues

Let us consider the observable $\widehat{\alpha}$ and its eigenstates (eigenvectors), each defined by the eigenvalues equation

$$\widehat{\alpha}|\eta\rangle = \alpha|\eta\rangle \tag{5.45}$$

with $|\eta\rangle$ the eigenstate (eigenket) fullfilling the equation with eigenvalue α (a scalar). We may assume that the observable $\widehat{\alpha}$ be characterized by a discrete (infinite) set of eigenvalues, i.e., a set of discrete variables $\boldsymbol{\lambda}_l$ fully defines each possible eigenstate and corresponding eigenvalue, hence providing

$$\widehat{\alpha}|\eta_l\rangle = \alpha_l|\eta_l\rangle \tag{5.46}$$

$$\langle\eta_l|\widehat{\alpha}|\eta_l\rangle = \alpha_l\langle\eta_l|\eta_l\rangle \tag{5.47}$$

which in case the eigenstates and the observable operator are expressed via any discrete orthonormal basis set, become usual matrix equations

$$\widetilde{\alpha}\boldsymbol{\eta}_l = \alpha_l\boldsymbol{\eta}_l \tag{5.48}$$

$$\left(\boldsymbol{\eta}_l^*\right)^T\widetilde{\alpha}\boldsymbol{\eta}_l = \alpha_l\left(\boldsymbol{\eta}_l^*\right)^T\boldsymbol{\eta}_l \tag{5.49}$$

where in Eqs. 5.47 and 5.49 $\langle\eta_l|\eta_l\rangle = \left(\boldsymbol{\eta}_l^*\right)^T\boldsymbol{\eta}_l = 1$. Note that when using the $\boldsymbol{\eta}_l$ eigenstates as the orthonormal basis set employed, the matrix $\widetilde{\alpha}$ becomes diagonal as following from Eq. 5.40.

It is interesting to consider another observable $\widehat{\beta}$, with discrete eigenvalues β_l, such that $\widehat{\alpha}\widehat{\beta} = \widehat{\beta}\widehat{\alpha}$ (i.e., the two observables commute), that is

$$\widehat{\alpha}\widehat{\beta}|\eta_l\rangle = \widehat{\beta}\widehat{\alpha}|\eta_l\rangle = \alpha_l\widehat{\beta}|\eta_l\rangle \tag{5.50}$$

readily providing

$$\widehat{\alpha}\left(\widehat{\beta}|\eta_l\rangle\right) = \alpha_l\left(\widehat{\beta}|\eta_l\rangle\right) \tag{5.51}$$

which must be fullfilled for whatever $|\eta_l\rangle$ eigenstate and hence, necessarily, $\widehat{\beta}|\eta_l\rangle$ must be proportional to $|\eta_l\rangle$, that is

$$\widehat{\beta}|\eta_l\rangle = \beta_l|\eta_l\rangle \tag{5.52}$$

Equation 5.52 simply states that when two observables commute then they must share one complete set of eigenstates (obviously when two observables share the eigenstates they always commute). In quantum mechanics the eigenstates of an observable are essential to connect the mathematical framework to the physics observed. In fact, the underlying idea is that every time an observable is measured the system collapses in one of the observable eigenstates, providing as measured observable value the corresponding eigenvalue. Therefore, given the fact that each observable eigenvalue provides a possible measured observable value, we must necessarily have that the eigenvalues of any operator representing a physical observable must be always real numbers. This condition readily implies that the observables must be always represented by Hermitian operators, which guarantee real eigenvalues as it follows considering the bra form of Eq. 5.46

$$\langle \eta_l | \widehat{\alpha}^\dagger | = \alpha_l^* \langle \eta_l | \tag{5.53}$$

clearly providing

$$\langle \eta_l | \widehat{\alpha}^\dagger | \eta_l \rangle = \alpha_l^* \langle \eta_l | \eta_l \rangle \tag{5.54}$$

Therefore, when the linear operator $\widehat{\alpha}$ is Hermitian, i.e., $\widehat{\alpha} = \widehat{\alpha}^\dagger$, from Eqs. 5.47 and 5.54 we have

$$\alpha_l = \alpha_l^* \tag{5.55}$$

clearly meaning that for Hermitian operators we only have real eigenvalues. We can investigate further the properties of the observables (Hermitian operators) by considering that from Eqs. 5.46 and 5.47 and Eqs. 5.53–5.55 we have

$$\langle \eta_{l'} | \widehat{\alpha} | \eta_l \rangle = \alpha_l \langle \eta_{l'} | \eta_l \rangle \tag{5.56}$$

$$\langle \eta_{l'} | \widehat{\alpha} | \eta_l \rangle = \alpha_{l'} \langle \eta_{l'} | \eta_l \rangle \tag{5.57}$$

and so by subtraction we obtain

$$(\alpha_l - \alpha_{l'}) \langle \eta_{l'} | \eta_l \rangle = 0 \tag{5.58}$$

The last equation clearly means that when $\alpha_l \neq \alpha_{l'}$ it necessarily follows $\langle \eta_{l'} | \eta_l \rangle = 0$, i.e., any couple of different eigenstates corresponding to non-identical eigenvalues must be orthogonal. Furthermore, considering that linear combinations of eigenstates with identical eigenvalues (degenerate eigenstates) always provide new eigenstates of the same eigenvalue, we can conclude that it is always possible to define at least one complete set of orthogonal eigenstates for any Hermitian operator, i.e., for any observable the corresponding eigenstates can be always considered as given by orthogonal vectors. Note that for discrete eigenvalue observables characterized by finite length eigenstates, the previous results imply that the corresponding eigenstates can be always given by orthonormal vectors.

It is worth to introduce now the expectation value of the observable $\widehat{\alpha}$ for a system in the quantum state $|a\rangle$ via $\langle a | \widehat{\alpha} | a \rangle$, providing when using the $\widehat{\alpha}$ orthonormal eigenstates as basis set

$$\langle a | \widehat{\alpha} | a \rangle = \left(\sum_{l'} a_{l'}^* \langle \eta_{l'} | \right) \widehat{\alpha} \left(\sum_l a_l | \eta_l \rangle \right) = \sum_l |a_l|^2 \alpha_l \tag{5.59}$$

In the last equation, when $|a\rangle$ is normalized (i.e., $\langle a|a\rangle = 1$), we obviously have $\sum_l |a_l|^2 = 1$ and hence considering the expectation value as simply the mean measure over an infinite set of equivalent observations consistently with the expectation value obtained when $|a\rangle$ is a normalized eigenstate, we must conclude that each $|a_l|^2 = a_l^* a_l$ provides the probability that a quantum system initially (before the measure process) in the normalized state $|a\rangle$ can be found (collapses) via a single measure of the observable $\widehat{\alpha}$ in the eigenstate $|\eta_l\rangle$ thus providing as measured observable value the eigenvalue α_l. Therefore, from

$$\langle a|\ |\eta_{l'}\rangle\langle\eta_{l'}|\ |a\rangle = \sum_{l''}\sum_{l} a_{l''}^* a_l \langle\eta_{l''}|\eta_{l'}\rangle\langle\eta_{l'}|\eta_l\rangle = a_{l'}^* a_{l'} \tag{5.60}$$

we can readily see that the Hermitian operator $|\eta_{l'}\rangle\langle\eta_{l'}|$, introduced in the previous section and diagonal in the $|\eta_l\rangle$ basis set, corresponds to the probability of finding the system in the $|\eta_{l'}\rangle$ eigenstate.

These results, as derived for observables with discrete eigenvalues and eigenvectors, can be extended to observables characterized by continuous or mixed eigenvalues and eigenvectors. In fact, all the previous equations of this section can be used even when the eigenvalues and eigenvectors are defined by either a set of continuous variables $\boldsymbol{x}$ or a combination of continuous and discrete variables (in both cases involving infinite length eigenstates) instead of only the discrete variables $\boldsymbol{\lambda}_l$. In such a case the continuous variables should be used in the definition of the eigenstates and integrals over such variables instead of summations are involved in the expressions corresponding to Eqs. 5.59 and 5.60. We can therefore assume that we may always define a proper basis set for the state space of a system by considering the set of orthogonal eigenstates shared by a group of commuting independent observables, where such eigenstates may define a discrete, a continuous or a mixed basis set according to the corresponding eigenvalues. In particular, when considering as basis set the eigenstates shared by a group of commuting observables with at least a subgroup of such observables involving continuous eigenvalues (continuous observables), we can use the continuous eigenvalues themselves to define the continuous variables $\boldsymbol{x}$ and the possible values of the discrete eigenvalues to define the discrete variables $\boldsymbol{\lambda}_l$. Within such a basis set the reference ket vectors $|\eta(\boldsymbol{x}_1)\rangle$ and $|\eta_{l_1}(\boldsymbol{x}_1)\rangle$ are the eigenstates corresponding for the continuous observables $\widehat{\boldsymbol{x}}$ to the eigenvalues $\boldsymbol{x}_1$ and for the discrete observables $\widehat{\boldsymbol{\lambda}}$ (observables with discrete eigenvalues), if present, to the eigenvalues $\boldsymbol{\lambda}_{l_1}$. Therefore, from Eqs. 5.17 and 5.21 we readily obtain that such eigenstates, each within its own basis set, are given by

$$|\eta(\boldsymbol{x}_1)\rangle = \delta(\boldsymbol{x} - \boldsymbol{x}_1) \tag{5.61}$$

$$|\eta_{l_1}(\boldsymbol{x}_1)\rangle = \boldsymbol{\eta}_{l_1}\delta(\boldsymbol{x} - \boldsymbol{x}_1) \tag{5.62}$$

$$\left[\boldsymbol{\eta}_{l_1}\right]_l = \delta_{l,l_1} \tag{5.63}$$

and hence from

$$\widehat{\boldsymbol{x}}|\eta(\boldsymbol{x}_1)\rangle = \boldsymbol{x}_1|\eta(\boldsymbol{x}_1)\rangle \tag{5.64}$$

$$\widehat{\boldsymbol{x}}|\eta_{l_1}(\boldsymbol{x}_1)\rangle = \boldsymbol{x}_1|\eta_{l_1}(\boldsymbol{x}_1)\rangle \tag{5.65}$$

we obtain that within such eigenstate basis sets the observables (vector operator) $\widehat{\boldsymbol{x}}$ must be simply given by the corresponding continuous eigenvalues (i.e., the vector of the eigenvalues), as it follows from the obvious mathematical relation (due to the definition of Dirac functions)

$$\boldsymbol{x}\delta(\boldsymbol{x} - \boldsymbol{x}_1) = \boldsymbol{x}_1\delta(\boldsymbol{x} - \boldsymbol{x}_1) \tag{5.66}$$

providing (when compared to Eqs. 5.64 and 5.65 and using Eqs. 5.61 and 5.62) $\widehat{\boldsymbol{x}} = \boldsymbol{x}$. Note that for the mixed basis set case (i.e., the representation obtained using the eigenstates of the $\widehat{\boldsymbol{x}}$ and $\widehat{\boldsymbol{\lambda}}$ observables as basis set) each discrete observable $\widehat{\lambda}_j$ $j = 1, n$ is expressed by a diagonal matrix whose non-zero diagonal elements are given by the eigenvalues $\lambda_{j,l}$ $l = 1, L_j$ where the upper limit L_j of the l index is defined by the properties of the discrete observables. Obviously when only discrete observables are considered to define the eigenstates to be used as basis set, at least for one jth observable $L_j \rightarrow \infty$. It must be noted that the relation $\widehat{\boldsymbol{x}} = \boldsymbol{x}$, valid when using the eigenstates of the $\widehat{\boldsymbol{x}}$ observables to define the basis set, implies that in the same basis set

$$\left(\widehat{x}_j\right)^M = \left(\widehat{x}_j\right)^{M-1} x_j = \left(\widehat{x}_j\right)^{M-2} x_j^2 = \cdots = x_j^M \tag{5.67}$$

$$e^{\beta\widehat{x}_j} = 1 + \beta\widehat{x}_j + \frac{\beta^2}{2}\left(\widehat{x}_j\right)^2 + \cdots = 1 + \beta x_j + \frac{\beta^2}{2}x_j^2 + \cdots = e^{\beta x_j} \tag{5.68}$$

$$\left(\widehat{x}_j\right)^{-M} = \left[\left(\widehat{x}_j\right)^{-1}\right]^M = \left(x_j^{-1}\right)^M = x_j^{-M} \tag{5.69}$$

$$\left(\widehat{x}_j\right)^{1/M} = x_j^{1/M} \tag{5.70}$$

where M is any natural number, β is an arbitrary constant, in Eq. 5.69 we used the relation $\left(\widehat{x}_j\right)^{-1} = x_j^{-1}$ which follows from

$$\left(\widehat{x}_j\right)^{-1}\widehat{x}_j = \left(\widehat{x}_j\right)^{-1} x_j = 1 \tag{5.71}$$

and Eq. 5.70 is simply due to

$$\left[\left(\widehat{x}_j\right)^{1/M}\right]^M = \widehat{x}_j = x_j \tag{5.72}$$

Equations 5.67–5.70, combined with trivial algebraic relations, readily show that any operator which may be expressed as a combination of powers and/or exponentials of $\widehat{\boldsymbol{x}} = (\widehat{x}_1\ \widehat{x}_2\ \ldots\ \widehat{x}_m)^T$ within the $\widehat{\boldsymbol{x}}$ eigenstate basis set must be expressed by the same overall function of the eigenvalues $\boldsymbol{x} = (x_1\ x_2\ \ldots\ x_m)^T$. More in general any arbitrary operator $\widehat{A}$ which is expressed by a combination of powers and/or exponentials of commuting operators $\widehat{\boldsymbol{\alpha}}$ (argument operators) necessarily commutes with each of such operators and hence the eigenstates of $\widehat{\boldsymbol{\alpha}}$ are also eigenstates of $\widehat{A}$, with the $\widehat{A}$ eigenvalues given by the same combination of powers and/or exponentials of the $\widehat{\boldsymbol{\alpha}}$ eigenvalues. Note that not any possible function of the $\widehat{\boldsymbol{x}}$ operators can be expressed by a combination of powers and/or exponentials and hence not any operator $\widehat{A} = F(\widehat{\boldsymbol{x}})$, within the $\widehat{\boldsymbol{x}}$ eigenstate basis set, can be necessarily given by $\widehat{A} = F(\widehat{\boldsymbol{x}}) = F(\boldsymbol{x})$. Only when fulfilling this last relation, i.e., the observables $\widehat{\boldsymbol{x}}$ eigenstates are also eigenstates of $F(\widehat{\boldsymbol{x}})$ with eigenvalues $F(\boldsymbol{x})$, $\widehat{A}$ can be considered an observable as its eigenvalues (its measured values) provide the proper expected relation with the eigenvalues of the argument observables (their measured values). Therefore, we will always assume that for any observable $F(\widehat{\boldsymbol{x}})$ even if not a combination of powers and/or exponentials of the $\widehat{\boldsymbol{x}}$ operators, when expressed in the $\widehat{\boldsymbol{x}}$ eigenstate basis set, we necessarily have $F(\widehat{\boldsymbol{x}}) = F(\boldsymbol{x})$.

5.4 The Coordinates-Spin Basis Set

In isolated physical–chemical quantum systems a typical example of a complete discrete basis set as defined by the eigenstates of commuting (discrete) observables is the set of eigenstates for the total energy, angular and linear momentum. A complete continuous basis set can be obtained for spinless particles simply by using the eigenstates of the system coordinates (i.e., commuting continuous observables) and a usual complete mixed basis set is obtained considering the eigenstates of the (commuting) coordinates and spin observables, i.e., the eigenstates of the coordinates (continuous observables) and spin operators (discrete observables). Note that the single particle spin operators $\widehat{s}_x, \widehat{s}_y, \widehat{s}_z$, corresponding to an intrinsic angular momentum with no classical analog and usually interpreted as resulting from the particle spinning rotation, provides an extra degree of freedom beyond the coordinates with discrete eigenstates defined as the eigenstates of the commuting square length and z-component of the particle intrinsic angular momentum.

When considering the continuous Cartesian coordinates $\boldsymbol{r}$ of all the system particles (i.e., the eigenvalues of the corresponding operators), the eigenstate defined by $\boldsymbol{r} = \boldsymbol{r}_1$ (as expressed within the coordinates eigenstate basis set, providing $\widehat{\boldsymbol{r}} = \boldsymbol{r}$ according to Eq. 5.66) is given by

$$|\eta(\boldsymbol{r}_1)\rangle = \delta(\boldsymbol{r} - \boldsymbol{r}_1) = \prod_{j=1}^{m} \delta(r_j - r_{j,1}) \tag{5.73}$$

where each $\delta(r_j - r_{j,1})$ Dirac function provides the eigenstate of the Cartesian coordinate operator $\widehat{r}_j = r_j$ with eigenvalue $r_{j,1}$ (i.e., the operator and eigenvalue of a single particle coordinate). For a system composed by N spin particles, by using the eigenstates of their spin observables and Cartesian coordinates to define the mixed basis set to be used, we can write for the eigenstate defined by a specific combination of the single particle spin eigenstates (the l_1 combination) and by $\boldsymbol{r} = \boldsymbol{r}_1$ (see Eqs. 5.62 and 5.63)

$$|\eta_{l_1}(\boldsymbol{r}_1)\rangle = \delta(\boldsymbol{r} - \boldsymbol{r}_1)\, \boldsymbol{\eta}_{l_1} = \prod_{j=1}^{3N} \delta(r_j - r_{j,1}) \prod_{k=1}^{N} \boldsymbol{\gamma}_{l_1}(k) \tag{5.74}$$

$$\langle\eta_{l_1}(\boldsymbol{r}_1)| = \delta(\boldsymbol{r} - \boldsymbol{r}_1)\, \boldsymbol{\eta}_{l_1}^T = \prod_{j=1}^{3N} \delta(r_j - r_{j,1}) \prod_{k=1}^{N} \boldsymbol{\gamma}_{l_1}^T(k) \tag{5.75}$$

where $\boldsymbol{\gamma}_{l_1}(k)$ is the kth particle basis set unit vector (defined by only one non-zero component equal to the unity) representing the kth particle spin observables $|\widehat{\boldsymbol{s}}_k|^2 = (\widehat{s}^2_{k,x} + \widehat{s}^2_{k,y} + \widehat{s}^2_{k,z})$, $\widehat{s}_{k,z}$ eigenstate (single particle spin eigenstate) involved in the full space basis set unit vector $\boldsymbol{\eta}_{l_1}$ (i.e., the l_1 spin combination eigenstate with components $\delta_{l,\underset{\sim}{l_1}}$) and from the definition of the matrix direct product for the $n \times n'$ $\widetilde{A}$ and $m \times m'$ $\widetilde{B}$ matrices

$$\widetilde{A} \bigotimes \widetilde{B} = \begin{bmatrix} a_{1,1} & a_{1,2} & \dots & a_{1,n'} \\ a_{2,1} & a_{2,1} & \dots & a_{2,n'} \\ \dots & \dots & \dots & \dots \\ a_{n,1} & a_{n,2} & \dots & a_{n,n'} \end{bmatrix} \bigotimes \begin{bmatrix} b_{1,1} & b_{1,2} & \dots & b_{1,m'} \\ b_{2,1} & b_{2,1} & \dots & b_{2,m'} \\ \dots & \dots & \dots & \dots \\ b_{m,1} & b_{m,2} & \dots & b_{m,m'} \end{bmatrix}$$
$$= \begin{bmatrix} a_{1,1}\widetilde{B} & a_{1,2}\widetilde{B} & \dots & a_{1,n'}\widetilde{B} \\ a_{2,1}\widetilde{B} & a_{2,1}\widetilde{B} & \dots & a_{2,n'}\widetilde{B} \\ \dots & \dots & \dots & \dots \\ a_{n,1}\widetilde{B} & a_{n,2}\widetilde{B} & \dots & a_{n,n'}\widetilde{B} \end{bmatrix} \tag{5.76}$$

with

$$a_{i,j}\widetilde{B} = \begin{bmatrix} a_{i,j}b_{1,1} & a_{i,j}b_{1,2} & \dots & a_{i,j}b_{1,m'} \\ a_{i,j}b_{2,1} & a_{i,j}b_{2,1} & \dots & a_{i,j}b_{2,m'} \\ \dots & \dots & \dots & \dots \\ a_{i,j}b_{m,1} & a_{i,j}b_{m,2} & \dots & a_{i,j}b_{m,m'} \end{bmatrix} \tag{5.77}$$

we have

$$\prod_{k=1}^{N} \boldsymbol{\gamma}_{l_1}(k) = \boldsymbol{\gamma}_{l_1}(1) \bigotimes \boldsymbol{\gamma}_{l_1}(2) \bigotimes \cdots \bigotimes \boldsymbol{\gamma}_{l_1}(N-1) \bigotimes \boldsymbol{\gamma}_{l_1}(N) = \boldsymbol{\eta}_{l_1} \tag{5.78}$$

where the dimension L of the full space basis set unit vector $\boldsymbol{\eta}_{l_1}$ is given by $L = \prod_{k=1}^{N} J_k$, with J_k the kth particle (finite) number of spin eigenstates, i.e., the dimension of each kth particle spin eigenstate unit vector $\boldsymbol{\gamma}(k)$. Note that for any $n \times n$ $\widetilde{A}$, $\widetilde{C}$ matrices and $m \times m$ $\widetilde{B}$, $\widetilde{D}$ matrices we always have

$$\left(\widetilde{A} \bigotimes \widetilde{B}\right)\left(\widetilde{C} \bigotimes \widetilde{D}\right) = \left(\widetilde{A}\widetilde{C}\right) \bigotimes \left(\widetilde{B}\widetilde{D}\right) \tag{5.79}$$

and within the basis set defined by the kth particle spin eigenstates each particle spin (vector) operator $\widehat{\boldsymbol{s}}_k$ is represented by three $J_k \times J_k$ matrices $(\widetilde{\varsigma}_{k,x}, \widetilde{\varsigma}_{k,y}, \widetilde{\varsigma}_{k,z})$ with clearly $\widetilde{\varsigma}_{k,z}$ and $\widetilde{\varsigma}^2_{k,x} + \widetilde{\varsigma}^2_{k,y} + \widetilde{\varsigma}^2_{k,z}$ diagonal matrices, e.g., for the kth particle corresponding to an electron we have

$$\widetilde{\varsigma}_{k,x} = \frac{\hbar}{2}\widetilde{\sigma}_x \tag{5.80}$$

$$\widetilde{\varsigma}_{k,y} = \frac{\hbar}{2}\widetilde{\sigma}_y \tag{5.81}$$

$$\widetilde{\varsigma}_{k,z} = \frac{\hbar}{2}\widetilde{\sigma}_z \tag{5.82}$$

with

$$\widetilde{\sigma}_x = \begin{bmatrix} 0 & 1 \\ 1 & 0 \end{bmatrix} \; \widetilde{\sigma}_y = \begin{bmatrix} 0 & -i \\ i & 0 \end{bmatrix} \; \widetilde{\sigma}_z = \begin{bmatrix} 1 & 0 \\ 0 & -1 \end{bmatrix} \tag{5.83}$$

the Pauli matrices.

To express the operator $\widehat{\boldsymbol{s}}_k$ within the full space basis set (via the matrices $\widetilde{s}_{k,x}, \widetilde{s}_{k,y}, \widetilde{s}_{k,z}$) we only have to use the direct product when considering for any other spin particle the corresponding $J \times J$ identity matrix

$$\widetilde{s}_{k,x} = \widetilde{I}_1 \bigotimes \widetilde{I}_2 \bigotimes \cdots \bigotimes \widetilde{\varsigma}_{k,x} \bigotimes \widetilde{I}_{k+1} \bigotimes \cdots \bigotimes \widetilde{I}_N \tag{5.84}$$

$$\widetilde{s}_{k,y} = \widetilde{I}_1 \bigotimes \widetilde{I}_2 \bigotimes \cdots \bigotimes \widetilde{\varsigma}_{k,y} \bigotimes \widetilde{I}_{k+1} \bigotimes \cdots \bigotimes \widetilde{I}_N \tag{5.85}$$

$$\widetilde{s}_{k,z} = \widetilde{I}_1 \bigotimes \widetilde{I}_2 \bigotimes \cdots \bigotimes \widetilde{\varsigma}_{k,z} \bigotimes \widetilde{I}_{k+1} \bigotimes \cdots \bigotimes \widetilde{I}_N \tag{5.86}$$

and thus any spin operator expression needed can be obtained by employing the previous equations (note that the direct product while fulfilling the associative and distributive properties cannot fulfill the commutative property).

It is evident that we can fully identify each coordinates-spin full space eigenstate of the N particles by the corresponding Cartesian coordinates eigenvalues $\boldsymbol{r}$ and the complete set of single particle spin eigenstates $\boldsymbol{\gamma} = \{\boldsymbol{\gamma}(1), \boldsymbol{\gamma}(2), \ldots, \boldsymbol{\gamma}(k), \ldots, \boldsymbol{\gamma}(N)\}$. Therefore, for any arbitrary and possibly time-dependent ket $|\Psi(t)\rangle$ its representation in the coordinates-spin eigenstate basis set (i.e., the vector wavefunction $\boldsymbol{\Psi}(\boldsymbol{r}, t)$) can be properly and very conveniently expressed via

$$\begin{aligned} \boldsymbol{\Psi}(\boldsymbol{r}, t) &= \sum_l \Psi_l(\boldsymbol{r}, t)\boldsymbol{\eta}_l = \sum_l \Psi_l(\boldsymbol{r}, t) \prod_{k=1}^{N} \boldsymbol{\gamma}_l(k) \\ &= \Psi(\boldsymbol{r}, \boldsymbol{\gamma}, t) \end{aligned} \tag{5.87}$$

where $\Psi(\boldsymbol{r}, \boldsymbol{\gamma}, t)$ is a compact notation representing the complete vector wave function via its generic component fully defined by the coordinates-spin variables $(\boldsymbol{r}, \boldsymbol{\gamma})$, i.e., $\Psi(\boldsymbol{r}, \boldsymbol{\gamma}, t) = \Psi_l(\boldsymbol{r}, t)$ for $\boldsymbol{\gamma} = \{\boldsymbol{\gamma}_l(1), \boldsymbol{\gamma}_l(2), \ldots, \boldsymbol{\gamma}_l(k), \ldots, \boldsymbol{\gamma}_l(N)\}$.

Reference

1. Dirac PAM (1958) The principles of quantum mechanics. Clarendon Press, Oxford

Index

A. Amadei and M. Aschi, *Statistical Mechanics for Chemical Thermodynamics and Kinetics*, https://doi.org/10.1007/978-3-031-77929-9

MIX
Papier aus verantwortungsvollen Quellen
Paper from responsible sources
FSC® C105338

If you have any concerns about our products,
you can contact us on
ProductSafety@springernature.com

In case Publisher is established outside the EU,
the EU authorized representative is:
Springer Nature Customer Service Center GmbH
Europaplatz 3, 69115 Heidelberg, Germany

Printed by Libri Plureos GmbH
in Hamburg, Germany